THE ILLUSTRATED GUIDE TO
PIGS

How to choose them – How to keep them

BLOOMSBURY

First published in Great Britain in 2011

Bloomsbury Publishing plc
36 Soho Square,
London
W1D 3QY

A CIP catalogue record for this book is available from the British Library.

ISBN 978-1-4081-4040-6

10 9 8 7 6 5 4 3 2 1

Commissioning editor: Nigel Redman
Project editor: Lisa Thomas
Design by Julie Dando at Fluke Art

Printed in China by C&C Offset Printing Co., Ltd

This book is produced using paper that is made from wood grown in managed sustainable forests. It is natural, renewable and recyclable. The logging and manufacturing processes conform to the environmental regulations of the country of origin.

The publishers are grateful to the following for permission to reproduce copyright material:

pp.8-9 *The Pig* © the estate of Roald Dahl. From Dirty Beasts (Jonathan Cape and Penguin Books Ltd). Reprinted with permission by David Higham Associates Ltd.
p.21 *A Pig is a Jolly Companion* © 1973, 2001 by Thomas Pynchon. From Gravity's Rainbow (Penguin Classics). Reprinted with permission by Melanie Jackson Agency, LLC.
p.35 *Pigs are Playful* © Charles Ghigna.
p.72 *The Pig Who Thinks in English* © Alan Reynolds.
p.150 *Any Part of Piggy* © the estate of Noel Coward. Reprinted with permission by Alan Brodie Representation.

Contents

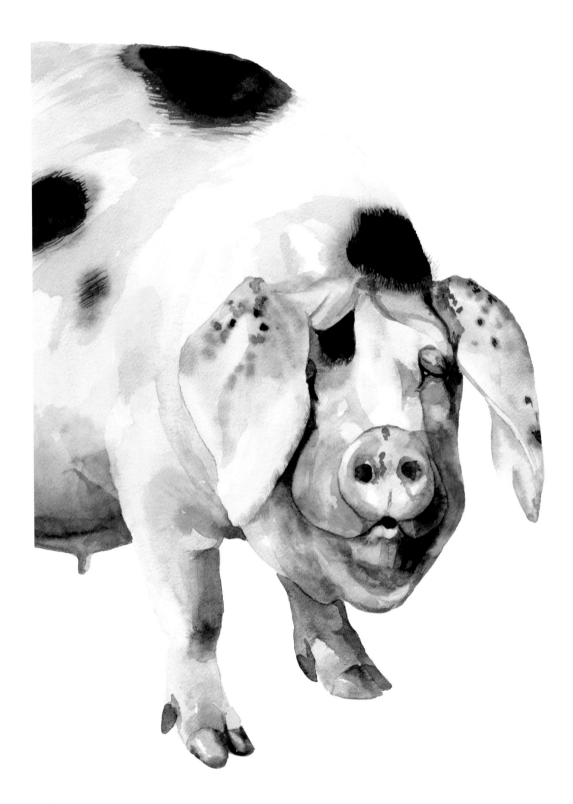

BUCKINGHAM PALACE

I am pleased to have been Patron of the Gloucestershire Old Spots Pig Breeders Club since November 2009 and indeed have raised a number of these historic pigs at my Gatcombe Park estate. So I am delighted to be able to provide a foreword for this book, a true celebration of these intelligent and endearing creatures, *An Illustrated Guide to Pigs*.

Known in Britain for hundreds of years, the Gloucestershire Old Spots was renowned for its prolific breeding, ease of management and the delicious pork and bacon that it produced. Yet during the 20th century, the GOS, along with many other traditional breeds, suffered a disastrous decline: by 1974, there were just 13 registered breeding boars. Dedicated work by enthusiasts has halted the GOS's decline, but several breeds, including the British Lop, Large Black and Middle White, remain endangered or vulnerable. Their loss would be a sad day for our agricultural heritage.

Today, we are all more conscious of our responsibility for the welfare of the creatures we rely on to feed us. Raising pigs slowly and in the traditional manner produces contented animals and better-tasting meat. By championing the unique and varied characteristics of our pigs, I hope that this book will inform a new generation of pig breeders and enthusiasts – and ensure the future survival of a wide range of breeds.

Anne

Introduction

For some reason the pig has historically attracted a great deal of bad press, most of it entirely unwarranted. The term 'sweating like a pig' is positively inaccurate, as pigs possess no sweat glands. And as one of the more intelligent animals it's hard to see why pigs give their name to the derogatory expression 'pig ignorant'? A 'dirty pig' is hard to find in the porcine world, as pigs never soil their beds if they can possibly help it. And as for 'pig ugly', beauty has always been in the eye of the beholder and many pig fanciers would hotly dispute this description.

The pig is a remarkably adaptable animal and famously every part of it can be used 'except the squeal'. The bristles are used for the finest hairbrushes and artists' paintbrushes, and once upon a time were used to make toothbrushes. The intestines provide skins for sausages; pigs' blood is turned into black pudding; while albumin, a protein in blood, is used in fixing pigment colours in cloth, clarifying liquors and making waterproof glues. Tusks and bones are used to make buttons and are an important component in bone china and bone meal in fertiliser.

Heart valves from specially bred pigs are successfully used in human heart transplants while pigs' pancreas glands provide a source of insulin for diabetics. As the hairs grow through the hide, pigskin leather is porous and breathable and suitable for items including gloves, shoes, footballs and saddles.

Pigs contribute all of the above – not to mention bacon, pork and lard. Today, different types of pigs are specially reared to produce pork or bacon – the former are more muscular and solid, the latter longer and leaner. In the past they were also specially bred for lard, once an important product used for cooking and as an essential lubricant, in the manufacture of soap, as lamp oil, in cosmetics and explosives. When newly developed synthetic lubricants and vegetable oils took over by the 1950s, the breeding of lard pigs declined.

Descendents of the Wild Boar (see pages 145–147), pigs were one of the very first animals to be tamed and domesticated by humans – perhaps as long as 9000 years ago. As omnivores, they can and do eat anything and everything, a fact that made their early domestication considerably easier for the farmer. Domestic swine are now found on every continent on earth except Antarctica.

By the Middle Ages in Europe, virtually every smallholder would have kept a 'cottagers pig', a sow that raised piglets over the summer. In the autumn when food became short the young pigs would be slaughtered and their meat salted to feed the family through the winter. Just a few boars would be kept and the 'boar walker' would come round in the spring. Each village had a swineherd whose responsibility was to take the pigs to forage for acorns or beechnuts. That all-encompassing document, the *Domesday Book* of 1086 even recorded the value of woodland according to how many pigs it was able to sustain.

Although individual pig keeping had declined following the great movement to urban areas that took place in the nineteenth century, during the Second World War many people were encouraged to keep a pig. As they could be economically fed on household waste, bins were put to hold for scraps for the pig. Today this rather random approach to nutrition would probably not be allowed, as government regulations require that all pigs must be vegetarian.

After the war, pig keeping turned into a commercial enterprise. By the end of the 1950s, large-scale pig keeping units raising thousands of animals were found everywhere, and the smallholder with a pig or two became a rarity.

Today, small-scale pig farming is gaining once more in popularity as people become more interested in where and how their meat is reared. Many breeds that were on the brink of extinction are gaining in numbers. The fact that people care more about how their food is produced also means that the huge indoor units, where pigs were kept in tiny stalls and never went outside, are also in decline and most commercial pig farmers are now proud to put 'outdoor reared' on their pork.

HOW PIGS WON THE WAR

Because of the pail,
    the *scraps* were saved,
Because of the scraps,
    the *pigs* were saved
Because of the pigs,
    the *rations* were saved,
Because of the rations,
    the *ships* were saved,
Because of the ships,
    the *island* was saved,
Because of the island,
    the *Empire* was saved,
And all because of
    the housewife's pail.

*Words from a Ministry of
Food advertisement 1941*

· · · · · · · · · · · · · · ·

This book is aimed both at people who are interested in raising their own pigs as well as those who simply love them in all their various forms. It gives an introduction to how to keep pigs, from deciding on the correct breed for your circumstances, to buying, transporting and settling them in. It covers aspects of feeding and looking after pigs including farrowing and dealing with ailments, followed by an illustrated guide to 38 popular breeds. The last section focuses on the slaughtering process, the various cuts of meat and explains how to cure bacon and ham and make sausages.

The Pig

In England once there lived a pig

A wonderfully clever pig.

To everybody it was plain

That Piggy had a massive brain.

He worked out sums inside his head,

There was no book he hadn't read.

He knew what made an airplane fly,

He knew how engines worked and why.

He knew all this, but in the end

One question drove him round the bend:

He simply couldn't puzzle out

What LIFE was really all about.

What was the reason for his birth?

Why was he placed upon this earth?

His giant brain went round and round.

Alas, no answer could be found.

Till suddenly one wondrous night.

All in a flash he saw the light.

He jumped up like a ballet dancer

And yelled, "By gum, I've got the answer!"

"They want my bacon slice by slice

"To sell at a tremendous price!

"They want my tender juicy chops

"To put in all the butcher's shops!

"They want my pork to make a roast

"And that's the part'll cost the most!

"They want my sausages in strings!

"They even want my chitterlings!

"The butcher's shop! The carving knife!

"That is the reason for my life!"

Such thoughts as these are not designed

To give a pig great peace of mind.

Next morning, in comes Farmer Bland,

A pail of pigswill in his hand,

And Piggy with a mighty roar,

Bashes the farmer to the floor...

Now comes the rather grizzly bit

So let's not make too much of it,

Except that you must *understand*

That Piggy did eat *Farmer Bland,*

He ate him up from head to toe,

Chewing the pieces nice and slow.

It took an hour to reach the feet,

Because there was so much to eat,

And when he finished, Pig, of course,

Felt absolutely no remorse.

Slowly he scratched his brainy head

And with a little smile, he said,

"I had a fairly powerful hunch

"That he might have me for his lunch.

"And so, because I feared the worst,

"I thought I'd better eat him *first."*

Roald Dahl

Why keep a pig?

There may be a number of reasons why you may be thinking of keeping a pig or two.

Raising for meat

If you've never kept a pig before it is highly recommended that you start by raising a couple of weaners to supply your freezer and see how you get on. This will entail buying in newly weaned pigs at around eight weeks old – these can be gilts or boars. You may have heard of 'boar taint' (a taste or smell that may affect the meat of uncastrated male pigs) but this will not be an issue if your animals go for slaughter at 20–24 weeks before they reach sexual maturity. An added bonus of boar weaners is that you won't be tempted to keep them to breed from which might be the case with gilts. If the weaners you bought are being sold for pork there may well be a good reason for this, such as poor conformation, and therefore they are probably not ideal for your foundation stock.

Breeding your own stock

Perhaps you want to keep a couple of purebred sows in order to produce weaners to sell yourself. If this is the case, make sure that there is a suitable boar nearby and buy the highest-quality registered stock that you can possibly afford. You may be able to buy an in-pig gilt which will overcome the problem of finding a boar for your first litter.

An unusual pet

Then again a characterful pet may be what you have in mind. Don't forget that a sweet little piglet may well grow to be 400kg (880lb) in weight and turn your garden into a ploughed field. A large black sow of my acquaintance, who lived happily free range in a farmyard, once became impatient when her meal was not delivered on time. She stuck her head through the catflap in the front door, removed the door with one shake of her head, and went to the kitchen to help herself to food! If you've set your heart on a pet pig you will need to acquire a pig walking licence. Contact your local Animal Health Office who will approve your route and issue you with an annual licence.

Choosing the right breed

Many breeds do stay smaller than others and some root less than others – but always make sure that you see the fully grown version of your chosen breed before you buy and if possible talk to someone who already owns the kind of pig you want. Pigs can live for around 20 years so a pet pig is not a short-term commitment. There is a vogue at the moment for mini or micro pigs – usually photographed looking achingly sweet in a teacup or some such – these are not a specific breed in themselves but have been bred from the runts of many different breeds specifically to be small. They are very expensive and you have no guarantee that they will stay small.

One other thing to consider, if your intention is to keep several different breeds of pig together; those with lop ears whose eyesight is constricted will have a distinct disadvantage over those with pricked ears, so it is best to restrict yourself to one or the other.

Never allow children in with pigs on their own – pigs can and do bite; they have surprisingly sharp teeth.

Terminology

Although a group of pigs is generally referred to as a 'herd', the old-fashioned word or collective noun is a 'sounder'. Here are some of the other words used to describe pigs of certain types or ages.

Baconer	A pig kept for a longer period than a porker or cutter to produce bacon (90–100kg; 200–220lb)
Boar	An uncastrated male pig
Barrow	A castrated male pig (USA)
Cutter	A pig raised for pork but older than a porker (64–82kg; 140–180lb)
Gilt	A female pig before she produces her first litter
Hog	Any domesticated pig in the USA
Piglet	Young pig until weaned
Porker	A pig raised for pork, generally five to six months old (55–62kg; 120–140lb)
Shoat	The name given to a young pig just after weaning (archaic)
Sow	A female pig that has produced her first litter
Weaner	A young pig that has been separated from its mother and is feeding itself – usually 6–10 weeks old

What to consider

Land

Pigs can be kept on very little land or a great deal of land. If they are being fattened for the freezer, containing them to some extent will speed up their rate of growth, but it will also encourage them to put on fat. Although they can be kept in a sty with a small run it makes sense to give them as large an area as you possibly can and give them room to root and follow their natural instincts.

Traditionally pigs were often kept in orchards but the orchard would need to be very large to avoid being totally dug over. Ideally the pigs would be let in only when the grass was high or windfalls were available. In a truly ideal situation you would have several runs and rotate the pigs, giving the land time to recover. Whatever system you use you will need to keep one area outside the sty that will always be ploughed as it is likely that the pigs will need to be kept off any other land during the winter.

PIGS AND HORSES

Horses have an innate fear of pigs – even the smell or sound may be enough to send them snorting and prancing in their field. Generally they soon get used to them and learn to enjoy their company, but there will be a few individuals that simply refuse to be ridden past pigs and never get over their fear.

Housing

All pigs need shelter – even in summer they must be able to get out of the sun. In winter they need somewhere warm and dry. There are many kinds of pig arks or stys on the market; you may be lucky enough to find one secondhand if you look in your local paper or smallholding magazine, or you could even look online. You can use a stable but remember that pigs are remarkably destructive, some more so than others, and whatever kind of shelter you use must be sturdy. The ark needn't be enormous but the pigs must have room to turn around and ample space to lie down. Put plenty of straw into the sty and the pigs will make their own beds. Site the sty out of any prevailing wind and if possible in a shady spot – also consider drainage.

He that makes himself dirt is trod on by the swine.
Proverb

Fencing

Pigs are ace escapees and if kept in a small area will spend a lot of their time thinking of ways to get out of it. Any fencing should be very strong and ideally backed up with a couple of strands of electric wire. Stock or pig netting should have an additional wire, preferably barbed, along the bottom to stop strong snouts from easing their way underneath. The more open space they have access to, the less likely it is that you will have to deal with escapees. It is surprising just how small a gap a large pig can squeeze through, though if they know their home and routine, once they have been out on an exciting adventure they may well decide to squeeze back in as tea time approaches. Don't feed pigs too near to a fence as if some of the food gets through to the other side this will encourage pushing and digging.

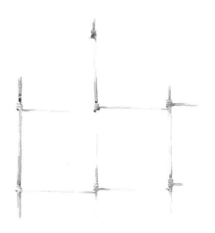

Electric fencing can also be used to divide a paddock – use a mains energiser unit rather than battery as the pigs will know at once if the battery goes flat and the mains unit will be considerably stronger. One other thing to consider is that if your fence is near a public footpath or right of way you should put up a clearly visible sign warning that the fence is electrified.

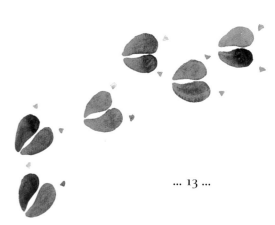

Water

Pigs drink a lot of water and are very good at tipping over any container. Ideally you should install an automatic water system but if this is impossible make sure that any trough is tied down or wedged in some way – a plastic container that fits inside a tyre or an old Belfast sink would be heavy enough to stay upright.

Pigs also need to be able to wallow. This is because they are unable to sweat and use this method to keep cool in summer and prevent sunburn. Wallowing and coating their skin with mud also keeps insects at bay and what is more, they enjoy it. A wallow is simply a hole in the ground that is filled with water daily.

AN EARLY FORM OF FOOTBALL?

And nowe in the winter, when
    men kill the fat swine
They get the bladder and blow it
    great and thin,
With many beans and peason
    put within:
It ratleth, soundeth, and shineth
    clere and fayre
While it is throwen and caste up
    in the ayre,
Each one contendeth and hath a
    great delite
With foote and with hands the
    bladder for to smite;
If it fall to grounde, they lifte it
    up agayne,
But this waye to labour they
    count in no payne.

Anon

A pig can drink up to 64 litres (14 gallons) of water a day and eat 1 ton of food a year – or 2.7kg (6lb) of food every day.

What to do next

Once you've made your decision and decided that you are going to keep pigs, before you go any further you must by law acquire a County Parish Holding number (CPH) for the land where the pigs will be kept.

This is a nine-digit number. The first two digits relate to your county, the next three relate to the parish and the last four digits are a unique number for you. You apply for a CPH from The Rural Payments Agency and they will supply you with your personal number.

You must then notify your local Animal Health Office (AHO) and give them your CPH number and they will issue a herd mark. Herd marks for pigs are one or two letters followed

SWELLFOOT

What! ye who grub
With filthy snouts my red potatoes up
In Allan's rushy bog? Who eat the oats
Up, from my cavalry in the Hebrides?
Who swill the hog-wash soup my cooks digest
From bones, and rags, and scraps of shoe-leather,
Which should be given to cleaner Pigs than you?

Percy Bysshe Shelley, Oedipus Tyrannus or Swellfoot the Tyrant

by four digits. You will need to have this mark imprinted on ear tags or slap-marked onto pigs that are leaving your holding. If you buy pigs at less than 12 months of age they can move with a temporary identifying mark and will only need to be tagged when they leave your premises.

Any movement of your pigs must be accompanied by an animal movement licence or AML2. This comes with four copies: the person you are buying the pigs from completes section A and C and keeps the yellow copy, the person transporting the pigs completes section B and keeps the blue copy, while the AML2 travels with the pigs. When they arrive you must complete section D and return it to your local authority.

Once your pigs arrive on your holding the holding will be under what is called a 20-day standstill. This means that no pigs may be moved off your holding for that length of time. All other animals on your property will also be on standstill for varying lengths of time.

All of these rules are set out in greater detail on the Department for Environment Food and Rural Affairs' (Defra) very comprehensive website.

Buying a pig

What to look for in a healthy pig

shiny bristles with no redness or flaky skin

Long straight back body well covered and not too fat

Clean warm ears correct for breed

Curly tail correctly set

Bright eyed and alert

Broad hams

12-15 evenly spaced teats

Cold moist snout not runny

Well sprung ribs

No sign of lameness

straight hocks

Acquiring stock

It is of course possible to just pop along to a local market or auction and pick up a couple of weaners, but the novice pig keeper will do far better to buy from a local breeder. It will be invaluable to have access to an experienced pig keeper who can offer advice and may even be able to provide an 'after sales service' if problems are encountered. The best time to buy weaners is in spring when there is summer ahead, plenty of grass growth and dry ground. Cross bred pigs are fine if your weaners are destined for the freezer and correct conformation and colouring are unimportant as long as the pigs appear healthy.

The actual lines of a pig (I mean a really fat pig) are among the loveliest and most luxuriant in nature; the pig has the same great curves, swift and yet heavy, which we see in rushing water or in a rolling cloud.

G K Chesterton

If you are buying foundation stock and intend to breed from them then you must take more care with your first purchase. Visit as many local breeders as you can, decide on the breed you want, join the relevant breed society and then buy the very best stock that you can afford. A pure bred pig should always be registered. If it is not you have no guarantee that the pig is what the breeder claims. Looks can deceive and some crosses produce progeny that take more strongly after either parent and appear to be that particular breed.

Other benefits of owning a pedigree pig are that you may qualify for an exemption during an outbreak of foot-and-mouth disease; you will be able to ask more for your own weaners and you will be helping to conserve that particular breed. Also bear in mind that there must be a boar of the right breed near enough to serve your sow – artificial insemination (AI) is a possibility but it is never as successful (see page 27).

It is possible to buy a gilt that is in-pig – if this is your choice you won't have to wait so long for your first litter – but it is a more expensive route to choose.

Bringing your stock home

Before collecting your new animals, make sure that all your fencing is secure, the ark or sty has a good quantity of straw, water is present and if possible you have a supply of whatever food your pigs have been fed on to date – sudden changes in diet can cause scouring (see page 31) so if your nuts are different to the ones that the pigs are used to, ask for a couple of days worth so that you can change over gradually.

Weaners can travel in a box in the back of the car as long as your journey is not over 40 miles – use plenty of newpaper or straw and make sure they aren't in direct sunlight – a metal dog cage is ideal. They can also travel with a temporary paintmark (see Identification below). Pigs larger than weaners must travel in a trailer and if they are over 12 months of age must by law have an eartag, tattoo or double slapmark bearing your Defra herdmark.

Once you get them home you will be under a 20-day standstill (see page 15) and will need to register your pigs with Defra. You do this by contacting your local AHO (Animal Health Office) who will ask you for your CPH number. Once your pigs are registered you will be issued with a herd mark. Herd marks for pigs are one or two letters followed by four digits.

Identification

You can identify your pig in a number of ways – by using an eartag, ear notch, tattoo or double slapmark.

- ❏ **Eartag** An eartag must be stamped or printed – not hand written. It should contain the letter UK followed by your herdmark. Tags used for pigs going to slaughter must be metal or plastic and able to withstand carcass processing. Tags used for movements between holdings can be plastic.
- ❏ **Tattoo** Your herdmark on the ear – the letters UK are not required.
- ❏ **Slapmark** A permanent ink herd mark that is applied to each shoulder.
- ❏ **Temporary mark** A painted mark such as a cross or circle.
- ❏ **Ear notch** This is a method for identifying coloured breeds where tattoos would not show. In the UK the British Saddleback has a specific system for notching and there is a different one for other coloured breeds.

If you want to register a pig as a pedigree it will need to be double tagged or have a tag in each ear. The British Pig Association (BPA) recommends notching the ear as well in case the tags fall out. If all the methods fail and it loses its identification, the pig can be re-registered.

Keeping records

You are required to keep a record of any movement of pigs on or off your holding within 36 hours. Once a year you must record the maximum number of pigs present on your holding. These records must be kept for six years after you stop keeping pigs.

You are also required to keep records of all medicines used – this includes wormers and vaccinations – and to keep these records for three years. Both these records must be available for inspection by the Local Authority.

If you are a member of a specific breed society you will be required to provide an annual record of all your registered stock with their identification numbers.

It makes sense if you intend to breed, to keep a careful record of pedigree, date of service, date litter is due and the number of piglets.

Pigspeak

Pigs make a wide range of sounds with a variety of meanings. They don't actually say 'oink' but make a noise more like 'groink' – but pig sounds are very difficult to spell! Below are descriptions of some of the sounds to help you understand what your pig may be saying:

Short sharp bark: You gave me a shock
Series of barks: I'm suspicious, there is something new here and I might bite
Loud sharp groinking bark: I'm threatened and may attack
Quiet continuous groinking: I'm totally content and probably rooting
Loud sudden squeal: I've touched the electric fence
Continuous week week week squeal: Here comes supper at last
Grrr rr rr – a bit like a quiet lion's roar: I'm in season
Breathy in-out he hon he hon he hon: I know you and I'm pleased to see you
Quiet quick groink groink groink: Sow to piglets; also used when suckling
Protective sounding barking hah hah hah: Sow finding out why piglet is squealing
Low nasal arf arf arf: Boar to sow – 'you're my kind of gal'

Young pigs grunt just as old pigs grunted before them
Danish proverb

A pig's whisper: a very short space of time; properly a grunt – which doesn't take long.
Brewer's Dictionary of Phrase & Fable

Handling and moving your pigs

Moving pigs to somewhere they have no wish to go, or think they have no wish to go, is when you discover that your charming friendly animal is stubborn. Those with lop ears that obstruct their eyesight, although docile, can be even more stubborn than a more lively type with prick ears. Also take into account that your pig, even at quite a young age, is going to be considerably stronger than you and no amount of pushing or pulling will work – guile is what's required.

Teaching your pigs to follow you

Whatever happens you will never get anywhere if you become agitated along with your pigs. Stay calm and if necessary enlist help. It pays hands down to have handled your pigs in advance. This doesn't mean just patting and scratching them although they will enjoy this and get to know you. By far the easiest method of moving them is to teach them to follow a bucket; when you bring their food walk round their pen. Every now and then move their feeding area so that they follow you and are rewarded. You can also practice using a slapboard and a stick or paddle. The slapboard is simply a bit of hardboard or plywood with a handhold cut at the top. It acts like a mobile wall on one side of the pig and the stick is used to encourage forward movement.

One major sticking point can be when the pig has to cross the point where an electric fence has been removed. It may refuse point blank, but to encourage it to cross, disguise the line with straw or shavings. Prepare your route so that the pig has only one way to go. You can build an elaborate passageway if the route is short or use flags and sheets to create one.

Getting pigs into a trailer

Loading your pig into a trailer can be another trial. You can try and get the pig used to it by putting it on one side of the pen and occasionally feeding it inside but this may not be practical. Try and make the loading area as small as possible and make sure the pig can't get underneath the trailer. One strange method that apparently works, according to expert pig keeper Andy Case, is to tie a rope to a back leg and when the pig has been manoeuvred on to the ramp, pull back as hard as possible. The pig will pull against you – then let go suddenly and hopefully the pig will run forwards into the trailer. You also need some assistants ready to push at the right moment.

Catching piglets

Picking up and catching piglets is one more skill to be learnt. The correct way is to grasp the piglet by its back thigh, if held hanging down in this way piglets seem to stay calmer than if picked up by their stomachs, when they will squeal ferociously. Hold the piglet by its back leg and support the stomach with your other hand. You'll soon find out that it is no good just chasing a piglet around – it is much quicker and more agile than you, and surprisingly slippery. Get it into a corner with the help of a slapboard, and possibly an assistant as well.

IN PRAISE OF A PIG

A pig is a jolly companion,
Boar, sow, barrow, or gilt –
A pig is a pal, who'll boost your morale,
Though mountains may topple and tilt.
When they've blackballed, bamboozled, and burned you,
When they've turned on you, Tory and Whig,
Though you may be thrown over by Tabby and Rover,
You'll never go wrong with a pig, a pig,
You'll never go wrong with a pig!

Thomas Pynchon, extract from *Gravity's Rainbow*

Feeding

Feeding any animal is never an exact science. Often it is something that comes with experience – you will learn to recognise if your pig is doing well, too well or not well enough, by eye. There will be many factors to take into account – does your pig have access to any grazing or forage? What size is your pig? Clearly the larger the pig the more it will need. Smaller breeds such as the Kune Kune will need considerably less than a Large Black for instance. But a Tamworth will also need more than a Saddleback as it is a more energetic breed. What age is your pig? How cold is the weather? In winter animals need to be fed extra as they are using energy keeping themselves warm. Is your sow in-pig or does she have a litter to suckle?

How to feel for condition

In order to be able to tell how much fat your pig is laying down, get into the habit of placing your hand over the backbone and ribs and develop 'feel'.

- ❏ **Emaciated**: bones are visible without the need to feel.
- ❏ **Thin**: bones can easily be felt when the hand is laid on the skin without pressure.
- ❏ **Ideal**: bones can be felt when a firm pressure is applied.
- ❏ **Fat**: bones can only be felt if fingers are pressed in.
- ❏ **Obese**: no bones can be felt at all.

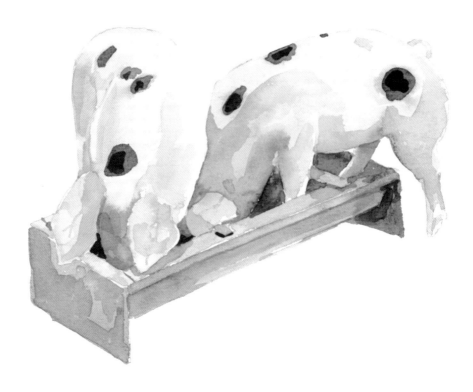

The golden rules of feeding – 'little and often' and 'introduce new foods gradually' apply to pigs. Their food must be broken down into at least two feeds – one huge meal would encourage scouring (see page 31), although having said that, piglets can be fed ad lib. Keep a record of what you feed until you feel confident that you have got the balance right.

Give to a pig when it grunts and a child when it cries and you will have a fine pig and a bad child.
Danish proverb

Varieties of food

There are several different types of ready made and balanced food that can be bought by the sack at your local feed merchant. It might be cheaper to feed your pigs 'straights': bags of single cereals such as barley and maize from which you create your own mix, adding vitamins and minerals of your choice. But it certainly won't be quicker and since the manufacturers of compound foods have spent time and money balancing their ingredients to produce the perfect nut or pellet it makes sense to use them.

There are various types of compound feed:

- **Creep pellets**: These are very small pellets, high in protein, that are fed ad lib to piglets in the 'creep' or part of the sty that the piglets can reach but the sow can't.
- **Grower pellets**: These are larger than creep pellets but smaller than sow nuts and with a higher protein content. These are usually gradually introduced two weeks before weaning and can be fed right up to pork weight.
- **Finisher pellets**: These are usually introduced at around 16 weeks and are fed for the final few weeks if you think your pigs are getting too fat.
- **Sow and weaner nuts**: These can be fed instead of grower pellets and are also suitable for in-pig gilts.
- **Pig nuts**: Fed to adult boars and sows that are not in-pig.

There is a difference between the pork of pigs that have been totally contained and those that have been allowed outside and taken exercise. The meat of the former will be pale and soft; while that of the latter will be firmer in texture and deeper in colour, even if they have both been fed the same diet.

A pig used to dirt turns its nose up at rice.
Japanese proverb

How much to feed

Take into account all the points made above and read the instructions on the feed sack of your chosen food. The following is a very rough guide for an average sized pig without access to any other forage. The amounts should be divided in half and fed twice a day. Even if you have a very large enclosure where your pigs can do a good deal of foraging for themselves, give them a feed at regular times each day to keep them tame and manageable, they will soon learn and you can even teach them to come to a whistle.

If the ground is exceptionally muddy, use a trough, otherwise throw the nuts on the ground for the pigs to forage around for them. This will take them longer and help to ward off boredom. A trough must be large enough for every pig to get to it at the same time or the smallest will constantly miss out. A round trough known as a Mexican hat or an oblong one with dividing bars (see page 22) are both ideal.

A good rule of thumb is 500g (1lb) for each month of age up to 2.7kg (6lb) for adult sows and boars. This is known as a maintenance diet. If you are rearing for pork this means you will have reached 1.8kg (4lb) by the time your pigs are 16 weeks and you should then continue to feed this amount until the pigs go to slaughter or they will put on too much fat. A lactating sow will need more depending on the number of piglets she is feeding – 2.5 to 3.5kg (6-8lb) and then a further pound for each piglet up to 7kg (15lb).

Every little helps said the sow
as she snapped at a gnat.
C. H. Spurgeon

British saddlebacks feeding from a 'Mexican hat'

How to approximate the weight of your pig

There may be occasions, such as knowing if your pig is ready for slaughter, when you want to know its weight and here is a simple way to find out. You need a full size tape measure then take the following steps.

1. Measure around the girth of your pig just behind the front legs (A).
2. Square the result.
3. Measure from the base of the ears to the base of the tail (B).
4. Multiply the first figure by the second.
5. If you are using inches, divide the result by 400 to obtain a weight in pounds.
6. If you are using metres multiply the result by 69.3 to obtain a weight in kilograms.

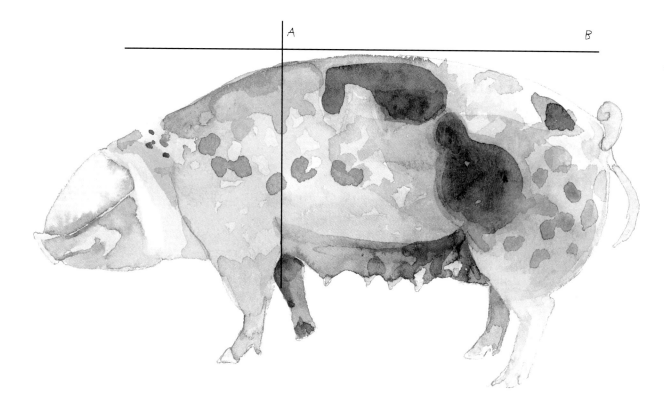

Other kinds of food

It is no longer legal in the UK and some other countries to feed your pigs with scraps or kitchen waste. This is to avoid the pigs ingesting bacteria that might result in foot-and-mouth disease. You may however, feed fruit and vegetables either from your garden or the local market and your pigs will certainly appreciate it – windfall apples are one example. Potatoes must be fed cooked. Milk and whey can also be fed but in the UK you need a licence from your local authority before you feed them.

An ungrateful man is like a hog under a tree eating acorns, but never looking up to see where they come from.

Timothy Dexter

A number of plants are poisonous to pigs, but as a rule unless the pigs are exceedingly hungry they will avoid them. Bracken is one, but as a general rule it is quite safe to keep pigs where bracken grows, and they may well do a good job of rooting it up, as long as they have access to plenty of fresh water. Ragwort, rhododendron, deadly nightshade and elder are all possible sources of poisoning and access to them should be avoided if possible.

It is a good idea to grow root crops such as turnips or Jerusalem artichokes if you have the space. If not, find a supply of fodder beet from a local farm. Feed merchants often have sacks of carrots for horses but pigs will be just as grateful.

Taking the organic route

There are strict regulations known as 'standards' that govern what organic farmers can or cannot do. Before you can call your pork organic you are required by law to acquire a certificate from an Independent Certification body. In the UK the Soil Association can provide this.

Among the regulations, pesticides are severely restricted, chemical fertilisers are banned, animal cruelty is prohibited and a truly free-range life promoted, the routine use of drugs, antibiotics and wormers is disallowed – instead the farmer must use preventative methods such as moving animals to fresh pasture and keeping a smaller herd size. Genetically modified (GM) ingredients in food are also banned under organic standards. Unless the pig you buy is organic you will have to wait for your first piglets before you call your pigs organic – a pig that is born 'unorganic' cannot become organic.

An inspector will come and inspect your farm or smallholding and talk through the necessary regulations if you want to go down this route; further information can be found on the Defra and Soil Association websites.

Boars

Should you keep a boar?

There are no two ways about it – boars are dangerous creatures. The occasional one may be a pussycat but as a rule even they must be treated with respect. If you go into a pen with a boar never take your eye off him, have a slapboard to protect yourself and don't stand right in front of him or close to the side of his head in range of his tusks. Stand by his shoulder if possible. The tusks can be trimmed but this will probably need to be done by a vet. Boars can be especially aggressive and unpredictable if they are near a sow in season. If you find you are the owner of an exceptionally aggressive boar, think hard about keeping him as he will pass this trait onto his progeny – in truth he is probably best turned into sausages.

It is not worth keeping a boar if you only have two or three sows although you could hire him out. The sows must always visit the boar as if you take him to them he will spend a good deal of time marking out his 'new' territory and neglect his duties. Remember that when the sow returns you will be under the mandatory 20-day standstill.

A boar will need very strong fencing at least 1.5m (5ft) high if he is near sows or he will simply bash his way through to get to them. Two adult males cannot be kept together; they may fight and severely injure each other. It goes without saying that if you do want to keep a boar you must get the very best animal money can buy.

Artificial insemination

The alternative to having a boar is artificial insemination (AI). There are many advantages to this system, such as no need for standstill, and being able to choose the very best boar available. The semen can be ordered by telephone (you can find suppliers on the internet or some breed societies can provide them) and will arrive by post in an insulated box the following day (avoid ordering on Fridays). Instructions will accompany the package and you will need a disposable catheter that is inserted into the sow's vulva.

It may take an experienced eye to know exactly when your sow or gilt is hogging (in season) although if you know your beast well it should be obvious. Her character may change and she may behave in a totally different way to normal. Or if your sow 'stands' while you press down on her back that is also a good sign. Another certainty is that she will come into season four to eight days after her piglets have been weaned.

Farrowing

If you do not have your own boar and choose to take your sow to one, try and take her just before her season is due which may be between eight and 36 hours. She will have to stay with the boar for at least two seasons or six weeks, to make sure that she doesn't 'return' or come back into season after 21 days, thereby showing that she is not in-pig. Mating isn't always a totally straightforward affair. The sow may take an instant dislike to the boar and attack him; your normally docile sow who walks calmly into your trailer may decide on this particular day that that is the very last thing she is going to do. You may even find you've got your dates wrong or by the time you arrive at the boar's home your sow is no longer hogging.

Caring for an in-pig sow

Hopefully all will be well and your sow will be in-pig when she returns. Do not over feed her, as she should not get too fat, so increase her rations only in the last three or so weeks to double what she has been getting. The gestation period for pigs is a neat 116 days, or three months, three weeks and three days. The sow's udder will begin to develop up to a month before she is due and harden with milk twenty-four hours before she actually gives birth.

Two or three days before she is due, she will start 'nesting' by pushing her bedding around and collecting grass and anything she can find to make a pile in the middle of her sty. If possible limit the amount of bedding she has – the less the better, as the new born piglets can get tangled in straw and are more likely to be squashed. Wood shavings on rubber matting are ideal, you can gradually re-introduce straw as the piglets grow and get stronger.

Getting ready for the birth

A couple of weeks before she is due, the sow should be separated from other pigs and wormed. Also check for external parasites and treat if necessary. She will be able to farrow quite happily in an ark; a special farrowing ark that has a rail running all around the edges so that the piglets are protected from squashing, is best. If the weather is cold a shed or stable may be more suitable as the piglets will need a heat lamp.

An area known as a creep can be partitioned off in such a way that the piglets can get in but the sow can't with the lamp suspended above. A special heat lamp can be bought from agricultural merchants and must be hung in such a way that neither the sow nor the piglets can come into contact with the wire. The temperature should be around 30–33 °C (85–95 °F) underneath the lamp.

What happens during the birth

As the birth approaches the sow will become restless, lying down and getting up and lying down again. Do not disturb her but stay nearby if possible. The sow will shake her tail as the first piglet appears.

The piglets are usually born nose first but occasionally one may be born backwards, in which case it may have mucous in its mouth – sweep this out with a finger and make sure it is breathing. If you suspect a piglet is not breathing, rubbing it vigorously with a towel and holding it upside down may help.

You do not have to be present at the birth, as the piglets will know exactly what to do. Within minutes of birth each will have found its way to a teat and be suckling, but it does help if you can prevent them being squashed which is a common occurrence. As births generally happen at night, unless you go out every hour to check it may well be that the birth takes place between your visits.

Dealing with newborn piglets

The birth can take from an hour and a half to eight hours and there may be half an hour between piglets. As each piglet is born, dry it with a towel and spray the umbilical cord with iodine (or dip it in a jam jar half full of iodine) and if the sow is calm put it onto a teat to suckle. If she is very restless it may be best to put the piglets into a box under the heat lamp, or with a hot water bottle, until she has finished. The womb is in two parts and the sow will get up and turn over to give birth to piglets on each side. As each side is emptied she will expel the afterbirth. The fact that there are also two placentas can come as a surprise to the novice pigkeeper.

The first-born piglets will get the front teat that has the most milk while the runt will be left with the one at the back. The piglets have favourite teats and will fight to keep ownership. The teats are in two rows, one above the other which enables a second piglet to lie on top of its brother or sister in order to reach the 'top deck'.

After a day or two the piglets will be much stronger, quicker and their squeal louder and although some sows seem immune to the agonised shriek of a piglet that is being lain on, a good mother will react and also take care when lying down.

The sow's ration should now be increased gradually depending on how many piglets she has – a good rule of thumb is an extra 500g (1lb) daily for every piglet up to 7.5kg (17lb) maximum.

Teeth clipping

During the first three days of life you may wish to clip the needle sharp teeth of the piglet with special clippers – this is done to protect the sow's udder and the piglets from injuring each other as if the litter is very large, fights may break out. With an average size litter however, teeth clipping should not be necessary.

Weaning

By eight weeks your piglets will be ready for weaning. They should have been sharing their mother's food for some time and she will also be tiring of allowing them to suckle. Take the sow away from the litter rather than the other way round, the piglets won't notice whereas if you try to move them they will all squeal in distress and everyone will get upset. If the sow is put out on new ground she will also be totally content.

Common ailments

Pigs are generally healthy creatures but there are a few health problems that may be encountered and the following are the most common. In the main you can treat your pigs yourself but if you feel your pig is not responding something more serious may be wrong and a vet must be consulted.

Whenever you feed your pig get into the habit of giving it an 'eye' – look it all over and see that the tail is curled – it may hang limply when the pig is standing still but should curl up when it walks (although this is not the case with Kune Kunes or Potbellies). Not eating is an obvious sign that something is wrong as is any sort of discharge or lameness.

You take a pig's temperature by inserting the thermometer in the rectum and 39°C (102.5°F) is normal.

❏ **Sunstroke** This is very common in pigs. You will see your pig shiver and stagger, or it may be reluctant to get up and appear sunburnt. Cool it down by sponging gently with cold water and make sure in hot weather it has plenty of shade and a wallow. A pale pink pig may need actual sunscreen applied if it is spending time outdoors in hot weather.

❏ **Worms** A pig with worms will lose condition and fail to put on weight; piglets may scour and develop potbellies. Pigs should be wormed at eight weeks and then regularly twice a year and worms should not be a problem.

❏ **Lice** Pigs love to rub and scratch and this is normal but if your pig seems to be scratching constantly, inspect it for lice. Lice are easily visible to the naked eye and look like tiny crabs – they particularly infest the region round the ears. Louse powder or an anti-parasitic wash will solve the problem or you can prevent it occurring in the first place by treating your pigs with an injection that will protect them from lice, mange and worms. (Bear in mind that injecting a pig is not a straightforward operation and will require guile and determination, not to mention assistance.)

❏ **Mange** Mange is also a parasite and manifests itself as scurfy red areas particularly on the head and leg and constant irritable scratching by the pig, perhaps with head shaking. Mange wash is available but preventing it with an anti-parasitic injection as explained above will protect your animals.

❏ **Scouring (diarrhoea)** Scours can be caused by a change of food, worms and food that the pig isn't used to – such as suddenly being given a large quantity of apples. Scouring should clear up by itself in a day or so. If it goes on it can cause dehydration and possibly death in piglets. Piglets often suffer scouring when they are weaned but this should only be temporary. A scouring powder that you add to water should stop the problem.

❏ **Hernias** Hernias, both umbilical and testicular, are fairly common in pigs and can be operated on, but as a rule do not cause a problem to the pig. In some cases they can disappear by themselves, but they can also be inherited. If several piglets from one litter develop hernias then you should consider using a different boar next time.

❏ **Lameness** If you find that your pig suddenly becomes lame, first inspect the trotter: it may have a stick or stone stuck between the toes. Check for cracks or sore patches and treat with an antibiotic spray such as Terramycin. If the pig is still lame after a day or so, call the vet as it will probably need treatment with penicillin. If you can find no injury suspect joint-ill which is a bacterial infection of the joints and will also require a visit from the vet.

Notifiable diseases

Foot-and-mouth This is a highly contagious disease that must be reported immediately if suspected. It is caused by a virus and the symptoms are lameness, salivating at the mouth and a high temperature. There may also be blisters on the snout, tongue and heels. There is no cure and if confirmed all your animals will be slaughtered.

Before you call the vet

It is no good calling the vet if your pig is out in a field and cannot be caught – this will only waste your time and his. If possible isolate the sick pig and bring it into somewhere enclosed where the vet can inspect it. Write down what the vet prescribes in your Medicine Book including the dosage and batch number. Make sure the vet tells you what he thinks is wrong so that you can prevent a repeat.

BREED PROFILES

Large Blacks

American Yorkshire

The most numerous pig in the United States

The pig that was to become known as the Large White in Britain retained its original name of Yorkshire when it was imported into the United States in the early nineteenth century. At first, the fact that the breed is slow to mature limited its popularity, but the American Yorkshire's ability to improve other breeds when crossed, soon made it one of the most popular pigs in the USA.

The American Yorkshire's docile nature make it ideal for breeding indoors in commercial units. In fact, if they are kept outside during the summer, plenty of shelter must be provided, as they are particularly prone to sunburn thanks to their pale skin. They are known to be excellent mothers, producing large litters with plenty of milk.

PIGS

Pigs are playful.
Pigs are pink.
Pigs are smarter
than you think.

Pigs are pudgy.
Pigs are plump.
Pigs can run
but never jump.

Pigs are loyal.
Pigs are true.
Pigs don't care
for barbecue.

Charles Ghigna

You can put wings on a pig, but you
don't make it an eagle.
William J Clinton

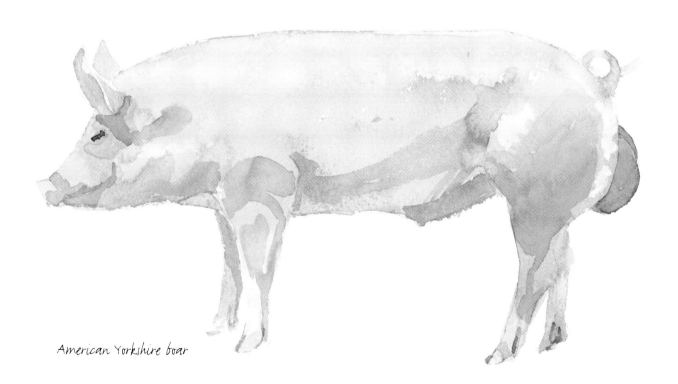

American Yorkshire boar

BACON AND EGGS

There was a young pig who, in bed,
Nightly slumbered with eggs on his head.
When the sun at its rise
Made him open his eyes
He enjoyed them for breakfast in bed.

Anon

American Yorkshire weaners

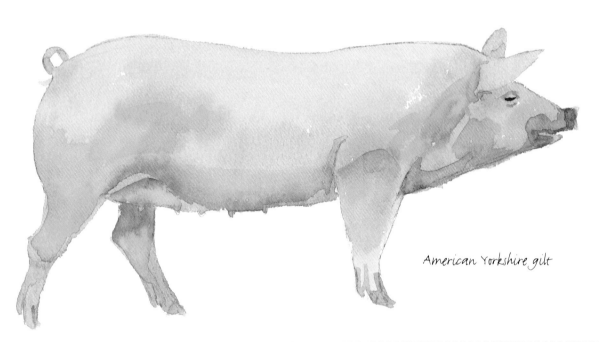

American Yorkshire gilt

ORIGIN	TYPE	SIZE	EARS	CHARACTER
USA	Bacon	Large	Erect	Docile

COLOUR
White coat over pink skin.

WHAT TO LOOK FOR – POSITIVE POINTS
Blemish free white coat. Long straight back. Small upright ears. Long dished face.

WHAT TO LOOK OUT FOR – NEGATIVE POINTS
Black spots on skin undesirable.

Bayeux

A Norman pig with a lazy streak

The Bayeux or Porc Bayeux, as its name suggests, originated in Normandy in the late nineteenth century. It is the result of crossing local Norman and British Berkshire pigs, producing a hardy animal that seemed to adjust well to intensive farming techniques. Thanks to its calm and endearing nature this is also an ideal breed for the smallholder. The Bayeux is an enthusiastic sleeper who never neglects a snooze in the sun – how else would they produce such delicious pork? They have a reputation for putting on weight fast, are prolific breeders and good mothers.

Sadly, during the Allied landings in Normandy in 1944, the breed was almost totally decimated. Very few animals survived and the Bayeux nearly died out altogether. However, some did endure and a few loyal farmers kept the breed alive. Today numbers are slowly increasing, thanks to a greater interest in small scale rearing.

Every autumn the city of Bayeux organises a Festival Gourmand du Cochon du Bayeux where there is a regional contest between the breeders – there is even a special song sung in praise of the pig to the tune of *La Bajocasse*.

SAILORS TURNED TO SWINE

The *Odyssey* is the epic story of the Greek hero Odysseus' ten years of adventure. It tells how, while returning from the Trojan War along the southern shore of the Mediterranean, the sorceress Circe enticed Odysseus' crew to come and have a meal with her. She then turned them into swine. But the god Hermes gave Odysseus a herb that made him immune to Circe's spells and he made her swear not to harm his crew. He drew his sword and forced her to turn them back into men, and Odysseus and his crew remained with her for a year.

Bayeux sow

Bayeux piglets

A pig's temperature is 39°C (102°F).
The pulse is 60 to 80 beats per minute.

Bayeux sow with piglets

ORIGIN	TYPE	SIZE	EARS	CHARACTER
France	Pork	Large	Lop	Calm and acquiescent

COLOUR
White with black spots.

WHAT TO LOOK FOR – POSITIVE POINTS
Rather short ears gently drooping. Elongated broad body with short legs.

WHAT TO LOOK OUT FOR – NEGATIVE POINTS
Predominance of black undesirable. Erect ears incorrect.

Bentheim Black Pied

A colourful pig with a compliant nature

Known in Germany as the Buntes Bentheimer Schwein, or 'colourful pig from Bentheim' or even Spotted German, the Bentheim Black Pied was originally developed from crosses between local north western German breeds and British Berkshires.

The numbers of this pig had been in gradual decline until in 2003 the Association for the Conservation of the Bentheim Black Pied Pig was set up. This body created a national herd book where previously the animals had been registered only in the regions or Länder. This has allowed reliable record keeping and a slow increase in the Bentheim Black's popularity.

This is a hardy, long-lived pig with a reputation for high fertility. It can produce on average around nine piglets per litter. Its superior flavoured meat makes it a popular choice for the small producer.

PIGS FOR GOOD LUCK

In the German language, there are a number of positive expressions that tie good luck to the pig. Someone who enjoys good fortune is told 'schwein haben' (to have pig). A person who picks the correct lottery numbers or wins a prize is said to be a 'lucky pig'. There is also a proverb that states: 'One who wants to make an impression buys a horse, one who wants to become wealthy breeds pigs.'

Bentheim Black Pied boar

BENTHEIM BLACK PIED

A pig has 15,000 taste buds, more than any animal including humans, and a very well developed sense of smell, ideal for their role as skilled truffle hunters.

Bentheim Black piglets

ORIGIN	TYPE	SIZE	EARS	CHARACTER
Germany	Pork and lard	Medium	Semi lop	Docile

COLOUR

White with black spots in grey rings.

WHAT TO LOOK FOR – POSITIVE POINTS

Gently dished rather long snout. Twelve working teats. Good level back. Medium sized ears that incline forwards.

WHAT TO LOOK OUT FOR – NEGATIVE POINTS

Predominance of black skin undesirable. Ears too large or upright.

Berkshire

A small pig with the chicest of coats

The Berkshire is the oldest recorded British pig, a smart looking animal with its four white feet and blaze. It is also known as the 'ladies' pig' as its small size makes it suitable for female pig handlers. Although generally placid and friendly, Berkshires can be mischievous, strong-willed characters.

The original Berkshire was a larger creature with a reddish coat and black or brown spots. The story goes that Oliver Cromwell's invading army discovered them when they entered Berkshire 350 years ago, and spoke of the excellent pork from local pigs. The addition of Chinese blood in the seventeenth and nineteenth centuries produced the animal we know today.

Berkshire gilt

Berkshire sow

Queen Victoria owned the first recorded boar, Ace of Spades, and kept a large herd at Windsor Castle. In fact a boar called Windsor Castle was the first import into the United States in 1841 – and he caused quite a stir as he weighed in at 1000lb (456kg).

THE EMPRESS OF BLANDINGS
Lord Emsworth was scion of the fictional Blandings Castle regularly featured in P. G. Wodehouse's 1920s comic novels. The Empress was his prize-winning Berkshire sow and Lord Emsworth was quite besotted with her, spending a great deal of his time on her care and diet. She was a frequent winner of the Fat Pig Class at the local Agricultural Show. The Empress appeared in many of the novels and there were a variety of devious plots and schemes that involved her kidnap.

Berkshire piglets

Berkshire sow

ORIGIN	TYPE	SIZE	EARS	CHARACTER
UK	Pork	Small	Prick	Placid but cheeky

COLOUR

Black with white on face, feet and tip of tail.

WHAT TO LOOK FOR – POSITIVE POINTS

Fine dished face with medium snout. Wide between the eyes. Fine sloping shoulders especially in females. Standing well on toes – a good walker. 12–14 working teats.

WHAT TO LOOK OUT FOR – NEGATIVE POINTS

White hair anywhere but as detailed above is incorrect. Crooked jaw. In-bent knees. Rose in coat.

Blanc de l'Ouest

A melange of a pig that enjoys wide, open spaces

The breed originates, as its name suggests, from the western areas of France: Brittany, Normandy and Pays de la Loire. It is the result of several breeds originally known in the Middle Ages, merging to become a single breed. Ultimately, it was a combination of the Normand of La Manche with the large dish-faced Craonnais – France's most renowned breed in the nineteenth century – that produced the Blanc de l'Ouest of today.

This large pig can stand up to 1 metre (32in) high at the shoulder and has a long body with a deep chest and good thick hams. It is happiest grazing free range for most of the year so is suited to the smallholder, rather than being kept in a large commercial unit. The sows make excellent mothers though the litters tend not to exceed eight piglets. Thanks to the pig's slow rate of growth, the meat has an excellent flavour and is much sought after by Parisien gourmets.

As with all pale-skinned pigs, sunburn can be a problem and provision of shade must be taken into consideration.

AUTUMN SACRIFICE

In Ancient Greece, it was believed that the winter season started in October. This was the time when the goddess Persephone was obliged to leave her mother Demeter (the goddess of fertility, the seasons, grain and abundance) to live in the Underworld with Hades. With her departure, everything ceased to grow. During the festival of Thesmophoria worshippers, who had raised piglets for the purpose, killed their animals and then buried them in a cave. The pigs' bodies would be dug up the next year and the remains mixed with the current seed to fertilise the new seed with the fruits of the old.

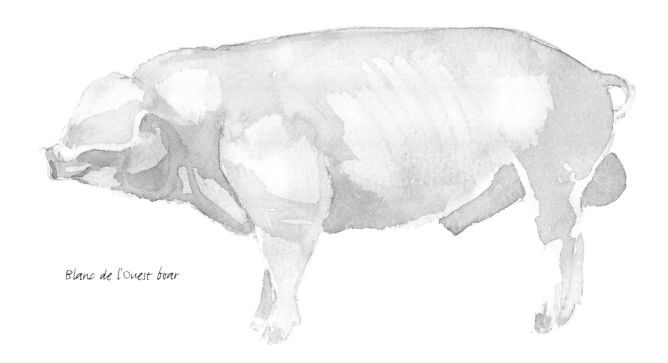

Blanc de l'Ouest boar

Blanc de l'Ouest piglets

Blanc de l'Ouest sow

ORIGIN	TYPE	SIZE	EARS	CHARACTER
France	Pork	Large	Lop	Peaceable

COLOUR

White.

WHAT TO LOOK FOR – POSITIVE POINTS

Broad forehead with dished snout. Long body with deep chest. Lop ears that meet at the end of the snout.

WHAT TO LOOK OUT FOR – NEGATIVE POINTS

Any pigmentation in the skin undesirable. Ears too short.

British Lop

A white pig with excessively long ears

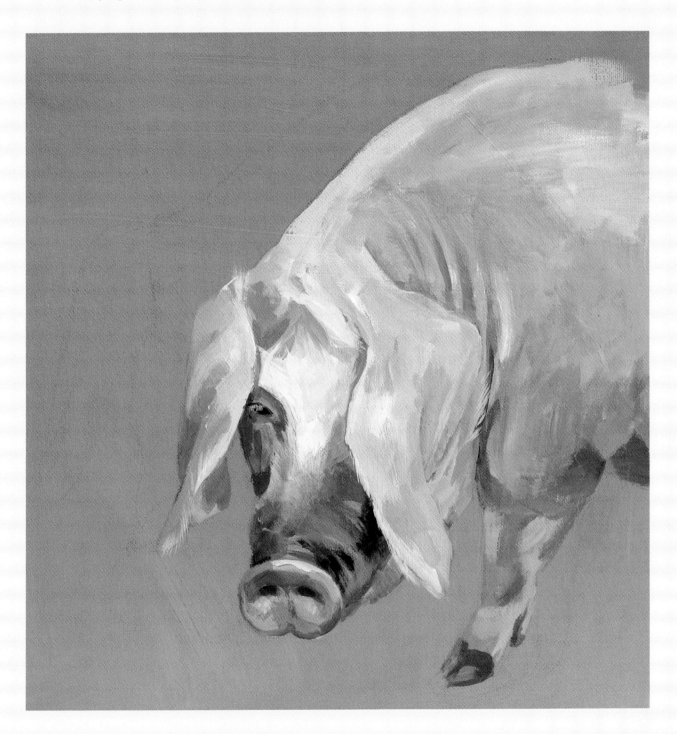

The British Lop is still found mainly in the West Country. It was originally known as the National Long White Lop Eared pig and is indeed very long in the body. When the Rare Breeds Survival Trust was established in 1973 the Lop was one of the first six breeds to be listed by the Society and this encouraged new interest in the pig.

The Lop is an easy to manage, hardy pig that finishes with a well-muscled lean carcass. The sows are excellent mothers and the boars make successful sires on other rare breeds producing well-fleshed pigs for the pork and cutter market.

One of nature's greatest delicacies is the black truffle and the most renowned come from the woodlands of south west France where they are still rooted out using the sensitive snouts of highly trained pigs. Sadly for the pigs, they rarely get to eat the priceless fungi they snuffle out, being rewarded instead with potatoes and apples.

British Lop

British Lop sow

One of the rarest of the native British breeds, the Lop looks rather similar to the Welsh (page 142) but according to the British Lop Society, their ears should resemble a dock leaf rather than a cabbage leaf and should be long and thin and reach to the tip of the nose – a fact that differentiates them from Landrace and Welsh pigs.

British Lop piglets

THE SWINEHERD PRINCE

The city of Bath was founded by a British king, Bladud, in 863 BC. Bladud had spent his youth in Athens, where he contracted leprosy. Knowing that a diseased prince could not become king, he left the royal palace and took a job as a swineherd in the Avon Valley. He drove his herd of pigs in search of acorns across the River Avon at Swineford. Like Bladud, the pigs were afflicted by leprosy, but when they rolled in the hot mud they found across the river, they were miraculously cured. Bladud then bathed in the springs and was similarly cured. He went home in triumph to be crowned king of the Britons. Bladud later founded a settlement at Bath, dedicating its curative powers to the Celtic goddess, Sul. When the Romans arrived 900 years later, they called the city Aquae Sulis, the Waters of Sul. And above the door of the baths, they set a carving of Prince Bladud casting acorns before his pigs.

ORIGIN	TYPE	SIZE	EARS	CHARACTER
UK	Pork or bacon	Large	Lop	Docile, amenable

COLOUR

White or pale pink.

WHAT TO LOOK FOR – POSITIVE POINTS

Ears should reach tip of nose. Long level back with straight underline. Tail set on high.

WHAT TO LOOK OUT FOR – NEGATIVE POINTS

Black or blue spots not acceptable. Rose in coat undesirable. Ears should not be too short.

British Saddleback

A handsome creature with a smart white belt

Originally there were two similarly marked breeds of Saddlebacks, the Wessex and the Essex. Both can trace their pedigrees back to 1918 when their societies were established and a herd book begun. The difference between the two types was minimal, having more to do with the area from which they hailed. Supposedly the Essex was thought to be a finer type of animal and known as the 'Gentleman's pig' whereas the Wessex was the 'Farmer's pig'. In 1967, the two breeds were amalgamated and became known as the British Saddleback, although a few breeders of pedigree Essex Saddleback pigs do still exist.

PIG IRON

This is a form of iron that is cast in oblong ingots. They are now known as pigs but were formerly referred to as sows. The term 'sow' is now applied to the main channel in which the molten liquid runs. The smaller branches that diverge from the sow are called pigs, and it is the iron from these that is called 'pig iron'.

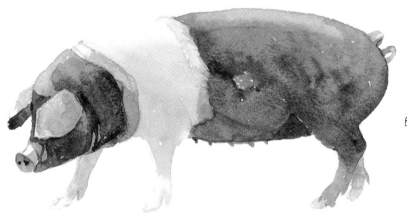

British saddleback gilt

British saddleback boar

The British Saddleback is a flexible animal that can be slaughtered at pork weight or raised to higher weights for bacon. The black pigmentation does not show on the meat itself. Thanks to its mainly black skin, sunburn is not a problem and it is, in fact, a very hardy breed. The sows make excellent, protective mothers of large litters and their laid back attitude to life makes them ideal pigs for the novice or smallholder.

British saddleback sow

British saddleback piglets

It brings good luck to carry
the wisdom tooth of a hog.
Proverb

ORIGIN	TYPE	SIZE	EARS	CHARACTER
UK	Pork and Bacon	Large	Lop	Easy-going

COLOUR

Black with white belt including the forelegs.

WHAT TO LOOK FOR – POSITIVE POINTS

Belt must be unbroken white. White tip to tail acceptable as are hind legs that are white up to the hock and a white nose. Ears should not obscure vision.

WHAT TO LOOK OUT FOR – NEGATIVE POINTS

White anywhere except as detailed above. Less than 12 sound and evenly spaced teats.

Chester White

An easily managed pig that likes the shade

Chester White

Asking a critic to name his favourite book is like asking a butcher to name his favourite pig.

John McCarthy

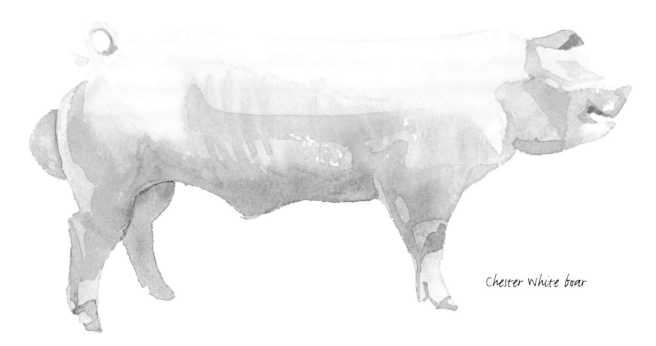

Chester White boar

Once known as the Chester County White, as they originated in Chester County, Pennsylvania, the Chester White Breed Association was established in 1884. Being pale in colour they suffer from sunburn and must have access to shade during the summer. The sows are sweet natured, good mothers producing large litters of 10 or 12 piglets. The boars however can be aggressive.

The Chester White is the most durable of the white breeds and versatile in that it can be farmed intensively in large units but will also do well free range on a smallholding. It is remarkable in that its daily weight gain can be as much as 600g (1.36lb) and for every 1.4kg (3lbs) of grain fed it will gain 450g (1lb) in weight.

Chester White and piglets

ORIGIN	TYPE	SIZE	EARS	CHARACTER
USA	Pork	Medium to large	Medium length lop	Tractable

COLOUR
White.

WHAT TO LOOK FOR – POSITIVE POINTS
Must have white skin and no coloured hair.

WHAT TO LOOK OUT FOR – NEGATIVE POINTS
Any coloured skin larger than a silver dollar would prevent registration.

Duroc

An American favourite with an adaptable coat

The origins of this large red pig are lost in the mists of time, but the name dates from 1823 when Isaac Frink from Saratoga County, New York bought a red boar from a litter bred by Harry Kelsey. Harry owned a famous trotting stallion called Duroc and Isaac called his future progeny after this horse. The Duroc was born.

Durocs are now the second most popular breed of pig in the USA. They weren't imported into Britain or Europe until the 1970s but have been gaining popularity ever since thanks to their innate suitability to the British climate. With a thick winter coat, Durocs can stand cold, damp winters but in the summer they moult and occasionally appear to be almost bald. They do not suffer from sunburn, rather their lack of a hairy coat allows them to stay cool and enjoy the heat.

THE LOUDEST SQUEAL

A pig can squeal at up to 130 decibels – to put that in perspective a jet taking off produces 140 decibels while the noise of a chainsaw is measured at 117 decibels.

Duroc boar

Duroc piglets

Durocs in commercial units are occasionally considered a rather aggressive breed and can be fiercely protective mothers. But if they are handled from birth, their naturally docile nature comes through and they are sociable and friendly. They have a reputation for converting feed into weight quickly and producing succulent and heavily muscled meat. The boars are frequently used as terminal sires for Landrace and Large Whites.

I learned long ago, never to wrestle with a pig. You get dirty, and besides, the pig likes it.

George Bernard Shaw

Duroc sow and piglets

ORIGIN	TYPE	SIZE	EARS	CHARACTER
USA	Pork	Medium to large	Semi-lop	Docile – quiet disposition

COLOUR

Skin – white or pink, at worst light grey. Hair – auburn. All shades of red from dark golden through cherry red to mahogany.

WHAT TO LOOK FOR – POSITIVE POINTS

Head small in proportion to size of body, face nicely dished. Short, thick and slightly arching neck.
Thick hair in winter, fine in summer.

WHAT TO LOOK OUT FOR – NEGATIVE POINTS

Ears standing erect, small cramped chest. Too small in size. No coarse, curly or white hair.

Gasçon

A black pig that produces top quality Jambon de Pays

This is a hardy, vigorous pig that tolerates heat and lives happily free range. Found in south-west France, in particular in the Haut Pyrenees, Garonne and Gers, it is similar to the Iberian (page 76). The Gasçon or Porc Gasçon has a rather long and exceptionally mobile snout with ears that are carried horizontally and are about half the length of the head. The tail is long and thick and ends in a luxuriant bunch of bristles.

The fact that the Gasçon is rather slow to mature means that its marbled meat has a superior taste and quality and produces firm bacon.

EATING THE CORRECT DIET

A cochon Gasçonnaise à dit
'I want spuds cooked in swill for
   my tea'
'Oh no' said the boar
'You must eat them raw.
For the better your bacon will be'.

Anon

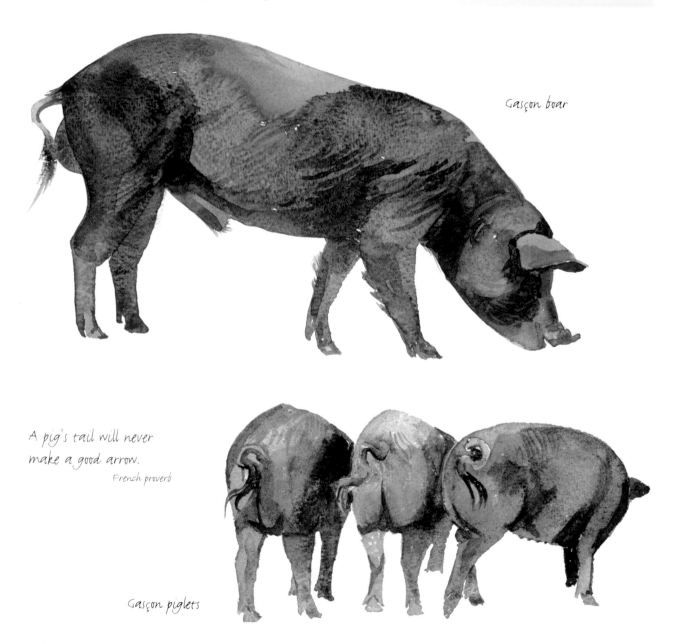

Gasçon boar

A pig's tail will never
make a good arrow.
French proverb

Gasçon piglets

Although numbers had dwindled rapidly from the 1950s to the 1980s, thanks to a number of rare breeds enthusiasts there is now a conservation programme in place for this breed and numbers are increasing.

Even a blind pig finds an acorn every once in a while.
Proverb

Gasçon sow

ORIGIN	TYPE	SIZE	EARS	CHARACTER
France	Pork	Large	Semi-lop	Active but good natured

COLOUR
Black.

WHAT TO LOOK FOR – POSITIVE POINTS
Ears should be horizontal, half the length of the head.

WHAT TO LOOK OUT FOR – NEGATIVE POINTS
Any white markings. Ears that are erect are incorrect.

Gloucestershire Old Spots

The world's oldest spotted pig

Gloucestershire Old Spots (GOS) pigs have been around for at least the last three hundred years and appear in many paintings of farm animals dating from the eighteenth century. They were once known as the 'orchard pig' as they were used to clean up orchard waste; legend has it that the black spots on their backs were bruises caused by falling apples.

In the 1920s Gloucestershire Old Spots were one of the commonest pigs around, particularly on the south side of the River Severn between Gloucester and Bristol, but fell out of favour as a result of poor breeding and the increase in intensive farming. The GOS is a pig well suited to the smallholder with a bit of land to spare as they are hardy and excellent foragers. The breed has recently enjoyed a comeback, although modern GOS pigs' coats are not as spotty as in former times. The sows make good mothers and can have litters at an age when other breeds are too old. When crossed with other white pigs, the spots disappear, so boars can be used to add hardiness and foraging skills to other breeds.

A few Gloucestershire Old Spots were imported into the USA in the nineteenth century and played a part in producing the Spotted Poland China but the breed never caught on in its own right, although there are a few loyal supporters who still keep them.

LEGENDARY BOAR HUNT

The Calydonian Boar or Hus Kalydonios, was a gigantic boar sent by the goddess Artemis to ravage the countryside of Calydon in Aetolia. She wished to punish King Oineus for neglecting her in his offering of the first fruits to the gods. The king summoned heroes from all over Greece to hunt down the beast and the famed Calydonian Boar Hunt which ensued was led by the king's own son Meleagros who struck the killing blow. The hero then awarded the skin-trophy to the huntress Atalanta as a prize for drawing first blood. She dedicated it to Artemis, hanging it from a tree in a sacred Arcadian grove.

Gloucestershire Old Spots sow

MONETARY PIGS

In the island of New Guinea, a person's wealth is judged in terms of the number of pigs owned; spending is measured in 'pig equivalents' and a person who does not eat pork is considered to be a heathen.

Gloucestershire Old Spots
sow with piglets

Gloucestershire Old Spots piglet

To get rid of a wart, peel an apple, rub it onto the wart and feed it to a pig. Another apparently reliable method is to rub the wart with a piece of pigskin once a day for a week.

ORIGIN	TYPE	SIZE	EARS	CHARACTER
UK	Pork/bacon/lard	Medium to large	Lop	Docile but stubborn

COLOUR
White with black spots and grey surrounds.

WHAT TO LOOK FOR – POSITIVE POINTS
Ears lop forward to nose but not longer. Slightly dished nose. Not less than one black spot – black should not predominate.

WHAT TO LOOK OUT FOR – NEGATIVE POINTS
Short thick elevated ears. Rose. Line of mane bristles. Sandy colour. Blue undertone not associated with a spot.

Guinea Hog

An all round small sized charmer unique to North America

This small black hairy hog originated in America's Deep South in the nineteenth century and is thought to have come over with slave ships that sailed from Guinea on the West African coast. It became a popular homesteader's pig thanks to its excellent foraging abilities that enabled it to fend almost entirely for itself. The pigs ate not only normal forage including nuts and apples but also enjoyed snakes, rats and any other small mammal that came their way.

This is a small pig and is nearly always black in colour, although very occasionally one is found with a rusty red coat. Some pigs have white markings on their feet and the tip of the snout and all have a hairy coat and upright ears. Thanks to their colouring they do not suffer from sunburn and enjoy heat but are equally at home in a cold climate.

THE GOLDEN BOAR

In Nordic mythology, Gullinbursti ('Gold-Bristle' or 'Gold-Mane') was the goddess Freyr's golden boar. The dwarves Brokkr and Sindri were challenged by the god Loki to make the boar for him. Sindri threw a pig's skin into a furnace while his brother worked the furnace and together they created Gullinbursti. The boar's shining fur is said to fill the sky, trees, and sea with light.

Guinea Hog sow

Fame is like a shaved pig with a greased tail, and it is only after it has slipped through the hands of some thousands, that some fellow, by mere chance, holds on to it!

Davy Crockett

Guinea Hog sow

Being lard pigs Guinea Hogs fatten up quickly on whatever they are fed and although smaller than many other breeds (though just the right size for one family) the quality of their meat is excellent. While it is still a rare breed in the USA, renewed interest in smallholding has engendered healthy interest in this charming animal that also makes an ideal pet.

Guinea Hog sow with piglets

ORIGIN	TYPE	SIZE	EARS	CHARACTER
US	Lard	Small	Large upright	Docile and friendly

COLOUR
Black but very occasionally rusty red. Minimal white feet and tip of nose.

WHAT TO LOOK FOR – POSITIVE POINTS
Compact-looking standing less than 65cm (30in) in height.

WHAT TO LOOK OUT FOR – NEGATIVE POINTS
Varies enormously in looks with long or short snout and long or short body – all acceptable, but there should not be more than a minimal amount of white hair.

Hampshire

A handsome pig with a smart white belt

Known originally as 'Thin Rinds' thanks to their thin skin, Hampshires take their name from their alleged port of origin. It is thought the breed was imported into the USA by sea from a port in Hampshire, England and is one of the earliest American breeds of hog that still exists today.

This belted pig is very popular in America thanks to its hardiness, vigour, prolificacy and foraging characteristics. They were particularly admired in the Corn Belt region during the early part of the twentieth century and have remained the pig of choice in many commercial farms ever since. This does not mean they are not an ideal breed for the smallholder.

The Hampshire hog is a handsome breed with a white belt that crosses the shoulders, encircles the body and covers the front legs. They are heavily muscled and produce lean meat and a carcass much admired by butchers. The sows are reliable mothers with good longevity but the boars can be aggressive.

Hampshire weaners

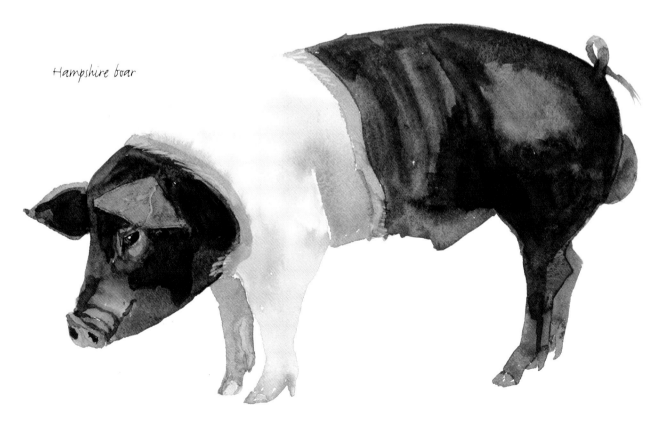

Hampshire boar

THE PIG WHO THINKS IN ENGLISH

The pig who thinks in English takes his ease
and taps his trotters daintily on tiles
that echo pleasantly while sun and breeze
bring pleasure to him, teasing out those smiles
he's famous for among his litter mates.
"What man is good for? There are many things.
They bring us dinner morning, noon, and night.
They track our pedigrees, record our weights
and wear, as we do, ear and nasal rings.
When alone, some like to warm our nights.
What's best? This fact, I think: it's really neat
how, if you close your eyes, they're good to eat."

Alan Reynolds

Pig in clover – someone who has any amount of money but doesn't know how to behave themselves as a gentleman.

Brewer's Dictionary of Phrase & Fable

Hampshire sow with piglets

ORIGIN	TYPE	SIZE	EARS	CHARACTER
USA	Pork	Large	Prick	Vigorous

COLOUR
Black with white belt across shoulders including front legs.

WHAT TO LOOK FOR – POSITIVE POINTS
Good straight back. Long deep sides. Slightly dished snout.

WHAT TO LOOK OUT FOR – NEGATIVE POINTS
Any white on the nose that goes over the rim. Any white above the hock on rear legs.

Hereford

An adaptable pig that suits any situation

This attractive, medium sized pig is found only in the United States. It was developed in Iowa and Nebraska during the 1920s from Duroc, Chester White and Poland China crosses. The colouring is reminiscent of the red and white colouring of Hereford cattle and this is how the pig got its name.

The colouring is important in this breed. In order for stock to be registered with the National Hereford Hog Record Association, pigs should have a primarily red coat, as deep as possible, with a white face and two or more white feet.

Herefords are adaptable beasts and thrive in many different systems and climates, but the fact that they are smaller than some breeds, are quiet and docile and known for their high feed efficiency, makes them ideal for the smallholder or first time pig keeper. The sows are good, caring mothers, producing large litters of colourful piglets.

True pig facts
- A pig has four toes but walks on two.
- A pig can run seven miles.
- Pigs are good swimmers.

Hereford gilts

Hereford piglets

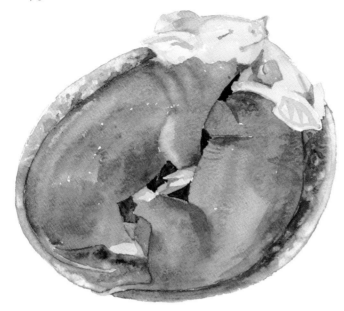

The film 'Babe', based on Dick King-Smith's children's novel about a pig that wants to be a sheepdog or 'sheep pig', premiered in 1995. During the filming, the piglets playing Babe had to be trained continuously because they could only be filmed at 16 to 18 weeks of age. In total 48 young pigs played the role.

KING NEPTUNE

King Neptune was a Hereford boar from West Frankfort, Illinois that a local farmer donated to be served at a fund raising pig roast during the Second World War. The local Navy recruiter decided that more money could be made if the pig was instead auctioned to raise funds for war bonds – in particular for money to build the battleship, *Illinois*. Pig and man travelled the length and breadth of Illinois but after each auction King Neptune was returned to be auctioned again. He became quite a celebrity and often appeared wearing a blue Naval blanket, crown and silver earrings. Over the course of his career he raised $19 million in war bonds. In 1946 he retired to a farm and a quieter life. Sadly, he died of pneumonia just two days short of his eighth birthday and was buried with military honours.

ORIGIN	TYPE	SIZE	EARS	CHARACTER
USA	Pork	Medium	Medium lop	Adaptable

COLOUR
Primarily red with white face and two or more white feet – deep red preferred.

WHAT TO LOOK FOR – POSITIVE POINTS
Slightly dished snout. Long neck with jowly appearance. Four white feet.

WHAT TO LOOK OUT FOR – NEGATIVE POINTS
White belt not acceptable. More than one-third white prevents registration. No rose.

Iberian

The king of all ham producers

The Iberian or Black Iberian is also known in Portugal as the Alentejano. This is a very ancient breed and it is thought that Phoenician traders brought the first pigs to the Iberian Peninsula. They came from the Eastern Mediterranean where they had interbred with wild boar.

Like their wild ancestors, Iberian pigs live naturally in the old oak forest or 'dehesa' where there are four types of oak including the holm and cork. It takes at least a hectare of dehesa to raise a single pig but it is the unique habitat that gives the Jambon Iberico its unique flavour. The main acorn harvest is taken from the holm oak from November to February, but the addition of the Spanish oak, gall oak and finally the cork oak, stretches the acorn season from September through to April.

Although the Iberian is a hairless pig, it is nonetheless a tough and hardy creature that by its very nature will do best outside – it has the ability to put on the fat that is so essential in the making of good quality dried ham. Iberians are not the best or most prolific of mothers and do not produce a huge quantity of milk.

OUTCAST PIGS
The Gaderene swine were the pigs into which Jesus cast demons that had possessed a madman and which as a result ran down a steep cliff into the sea and were killed. From this story, the term 'Gadarene' has come to refer to involvement in a headlong or potentially disastrous rush to do something.

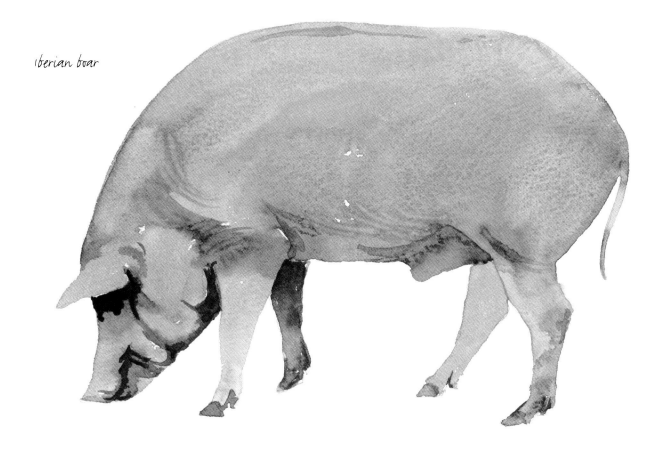

Iberian boar

A pig bought on credit is forever grunting.

Spanish proverb

Iberian gilts

Red Iberian

ORIGIN	TYPE	SIZE	EARS	CHARACTER
Spain	Lard and bacon	Large	Lop	Independent

COLOUR
Black but can also be dark grey and russet red.

WHAT TO LOOK FOR – POSITIVE POINTS
Sleek hairless skin. Long nose.

WHAT TO LOOK OUT FOR – NEGATIVE POINTS
Any white skin or coarse bristles are undesirable.

Iron Age

A wily creature that is hard to contain

The Iron Age pig is actually a hybrid originally created in the 1970s for *Living in the Past*, a BBC television programme about life in an Iron Age village. A Tamworth was crossed with a Wild Boar from London Zoo to produce the kind of pig that historians believed would have been around at that time.

Although tamer than Wild Boar, the Iron Age is still a fearsome looking beast with upright ears and a long snout. It also has a broader than normal ribcage and is well versed in every kind of escapology. This is a pig that must live free range, preferably in a wooded environment, though they will take to pasture grazing. It is also one of the most destructive rooters and any housing must have a concrete floor or it will simply be destroyed. It goes without saying that fencing must also be extremely robust.

THE BOAR-GOD

Varaha is the third Avatar of the Hindu god Vishnu and takes the form of a boar. He appeared in order to defeat Hiranyaksha, a demon who had taken the Earth and carried it to the bottom of the ocean. The battle between Varaha and Hiranyaksha is believed to have lasted for a thousand years, with Varaha the eventual winner. He carried the Earth out of the ocean between his tusks and restored it to its rightful place in the universe.

Iron Age sow

The Iron Age is prized for its exceptionally lean and gamey meat and is now produced for the speciality meat market. The piglets are stripey like Wild Boar piglets but lose their stripes as they mature and although litter sizes are generally small the mothers are particularly protective.

There is a saying that pigs can see the wind – if a pig is seen running with straw in its mouth it means that a storm is approaching.

Iron Age piglet

Iron Age sow with piglets

ORIGIN	TYPE	SIZE	EARS	CHARACTER
UK	Pork	Medium	Prick	Wild and crafty

COLOUR

Varying from rusty red to black, most commonly brown and similar to a wild boar.

WHAT TO LOOK FOR – POSITIVE POINTS

Broad ribcage. Upright ears and long snout.

WHAT TO LOOK OUT FOR – NEGATIVE POINTS

Less than 10 working teats.

Kune Kune

The cartoon character of the pig world

Although the Kune Kunes (pronounced *cooney cooney*) that are found today come from New Zealand – the word means 'fat and round' in Maori – they did not originate there as New Zealand has no indigenous land mammals. Pigs with 'tassels' are found in Polynesia and they may have made their way there by a variety of different routes.

Regardless of its origins, today the Kune Kune is a charming little beast no more than 60 to 75cm (24–30in) tall and weighs around 110kg (240lb). It is also covered in short, long or curly hair and can have ears ranging from prick to full lop. All Kune Kunes must have two tassels hanging under their chin like a goat. Although unusual, this feature is not unique – the American Red Wattle (page 127) also has them.

Kune Kune weaners at feed bin

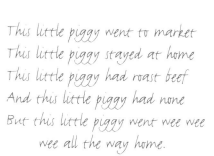

This little piggy went to market
This little piggy stayed at home
This little piggy had roast beef
And this little piggy had none
But this little piggy went wee wee
    wee all the way home.

Old nursery rhyme

Kune Kune sow

They come in a wide variety of colours and spots from ginger through white to black – with this breed anything goes. There can be fashions in the colour of this breed – the Maoris preferred black but in the UK there is a vogue for spotted. The British Kune Kune Society points out that breeders must be careful to breed all types and colours to preserve as much of the gene pool as possible.

Kune Kune piglet

Being small and docile, if you are looking for a pet pig this is the one for you. The Kune Kune won't cut up the land quite as badly as larger breeds, enjoys grazing and being smaller will be easier to cope with than larger more boisterous breeds.

sleeping Kune Kune piglets

ORIGIN	TYPE	SIZE	EARS	CHARACTER
New Zealand	Pet	Small	Prick or lop	Placid and friendly

COLOUR
May be any colour.

WHAT TO LOOK FOR – POSITIVE POINTS
Broad dished face with short to medium snout. Two tassels. Legs able to support the size.

WHAT TO LOOK OUT FOR – NEGATIVE POINTS
The vision should not be obstructed by too large ears. Should not be too fat and unable to run.

Lacombe

An all-round Canadian favourite

Alberta in Canada is home to the Lacombe, a pig specifically bred for its rapid weight gain, docility and large litter size. The foundation stock was a crossing of Berkshire sows with Landrace Chester White boars and this produced a superb creature that has excellent feed efficiency and is fast-growing. The breeders' main aim was to produce a terminal sire for crossing with the Yorkshire, the most popular breed in Canada.

Known for its docility, this is a pig that will do well if confined but is also hardy and can live free range. The sows are good mothers that give birth to large litters that grow fast and produce first class pork.

> Tom Tom the piper's son
> Stole a pig and away he ran,
> The pig was eat and Tom was beat
> And Tom went roaring down the street.
>
> *Old nursery rhyme*

The saying, 'Don't buy a pig in a poke' dates from seventeenth century England. At market it was a common trick to try to con unsuspecting buyers by selling them an animal unseen in a sack. So a person who thought he was buying a suckling pig might find when he opened the poke (sack), that it was in fact a cat and that he had, 'let the cat out of the bag'.

Lacombe gilt

Lacombe weaners

The largest litter of piglets ever recorded comprised 37 piglets of which 36 were born alive and 33 eventually survived.

Never eat more than you can carry
Miss Piggy

Lacombe piglet

ORIGIN	TYPE	SIZE	EARS	CHARACTER
Canada	Pork	Medium	Lop	Docile

COLOUR
White.

WHAT TO LOOK FOR – POSITIVE POINTS
Long body with rather short legs. Fine hams.

WHAT TO LOOK OUT FOR – NEGATIVE POINTS
Any black hair undesirable. Less than 12 working teats.

Landrace

The quintessential pig

The term Landrace literally means 'national breed' or 'breed of the land'. The USA and many European countries have their own Landrace breed – all of which are very similar. Danish bacon is invariably supplied by this breed, which was first imported into the USA in 1934. The first Landrace to reach Britain came from Sweden in 1949

The Landrace is a long, lean-bodied pig with lop ears and fine white hair. Its docile and submissive character has made it ideal for intensive farming and it is usually found in commercial indoor systems. The breed's strength is the speed at which it grows and its ability to improve other breeds. Sows produce large litters and are exceptionally heavy milkers, which enables the piglets to grow faster than in most other breeds.

Although docile, the Landrace is not a hardy breed and will suffer from sunburn outside – it might not be first choice for the smallholder.

The first pig to fly in an aeroplane was taken up for a joyride in a biplane by Lord Brabazon, holder of the first pilot's licence in Britain.

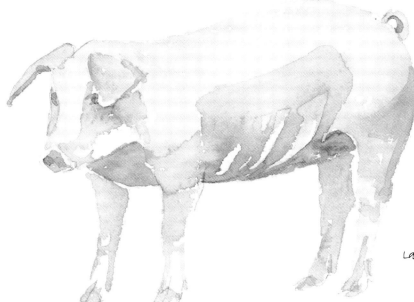

Landrace weaner

THE ADVENTURES OF TIRPITZ
During the First World War, pigs were often kept on board warships to supply fresh meat. Tirpitz was on board the German vessel *SMS Dresden* when the British warship, *HMS Glasgow* attacked her off the coast of South America. As the ship sank, the pig made his way onto the deck and then swam clear of the sinking ship where he was seen and rescued by the crew of the *Glasgow*. They named him 'Tirpitz' after the German Admiral and awarded him the Iron Cross for remaining with his ship. The pig lived on board the *Glasgow* for a year and when the ship returned home he was adopted by the Petty Officer who had first seen him in the sea. Tirpitz lived at the Whale Island Gunner School in Portsmouth until he was auctioned off for charity in 1919, raising £1785 for the British Red Cross. Tirpitz's head was mounted and is still to be seen in the Imperial War Museum. When *HMS Glasgow* was rebuilt, a pair of silver carvers made from Tirpitz's trotters was bequeathed to the ship.

LANDRACE

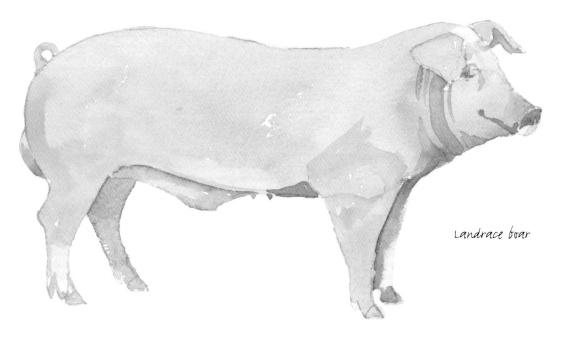

Landrace boar

Landrace sow

ORIGIN	TYPE	SIZE	EARS	CHARACTER
Europe	Bacon	Large	Lop	Submissive

COLOUR
White.

WHAT TO LOOK FOR – POSITIVE POINTS
Long lean body. Slightly arched back.

WHAT TO LOOK OUT FOR – NEGATIVE POINTS
Hair of any colour other than white. Upright ears. Black spots in the skin.

Large Black

Britain's only all-black pig

The Large Black was supposedly developed from Chinese pigs that came off merchant ships in south-west England and mated with the local swine. In some parts of the world the Large Black is still known as the Cornish Black. The Large Black Society was founded in 1889 and encouraged the immense popularity and spread of this breed; eventually it was found in over 30 countries.

However the rise in popularity of white pigs in commercial units led to the decline of the Large Black and it is now classed as a rare breed.

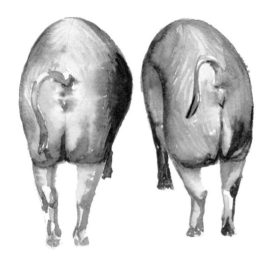

Large Blacks

A SWINEHERD'S VISION

During the eighth century AD, a swineherd called Eoves saw a vision of the Virgin Mary while searching for a stray pig in an area of Worcestershire known as Lomme. He went to see Ecgwin, the Bishop of Worcester, and reported his experience. When Ecgwin returned with him, he too saw the vision of Mary. In 709 Ecgwin founded an Abbey on the site of the vision and became its first Abbot. The area became known as Eoveshomme (later Evesham). By the time the abbey was dissolved by Henry VIII in the sixteenth century, it had become the third most important in England.

Large Black gilt

This is a very large breed of pig, but its lop ears, which restrict its vision make it somewhat easier to manage than other pigs of similar size. It also benefits from a calm temperament that makes it a good beginner's pig. The Large Black does not suffer from sunburn thanks to its black skin. This is a pig that can get a good deal of its food by foraging and the fact that it is slow to mature is reflected in the quality of the meat. Although they produce good quantities of milk, Large Black sows can be careless mothers prone to lying on their piglets.

One disadvantage of being a hog is that at any moment some blundering fool may try to make a silk purse out of your wife's ear.
J B Morton

Large Black sow and piglets

ORIGIN	TYPE	SIZE	EARS	CHARACTER
UK	Pork and bacon	Large	Lop	Temperate

COLOUR
Black.

WHAT TO LOOK FOR – POSITIVE POINTS
Long thin ears well-inclined over the face. Very long, strong back. Blue-black skin with fine black silky hair.

WHAT TO LOOK OUT FOR – NEGATIVE POINTS
A narrow forehead, dished or undershot lower jaw. Ears should not be 'cabbage leaved' and the coat should not be curly or coarse.

Large White

The World's favourite pig

The Large White is a finer version of the coarse old Yorkshire pig and was originally developed as an outdoor breed. However it was found to do extremely well indoors as well and is very popular with commercial pig farmers, thanks to its ability to cross and improve other breeds, and stamp uniformity and quality on their progeny.

This is a hardy and versatile pig with a sensible temperament that produces extremely large litters and the milk to go with them. Huge numbers of Large Whites have been exported from the UK all over the globe and today it truly can be considered to be the world's favourite pig.

PIG DICTATOR

In George Orwell's *Animal Farm*, Napoleon started life as an ordinary farm pig but, following the animals' rising against their human masters, he became the president of Animal Farm, eventually taking on the attributes of his former masters to become a violent and corrupt dictator. Orwell apparently based the character on Josef Stalin but named him after Napoleon Bonaparte.

I understand the inventor of the bagpipes was inspired when he saw a man carrying an indignant asthmatic pig under his arm. Unfortunately the manmade sound never equalled the purity of the sound achieved by the pig.

Alfred Hitchcock.

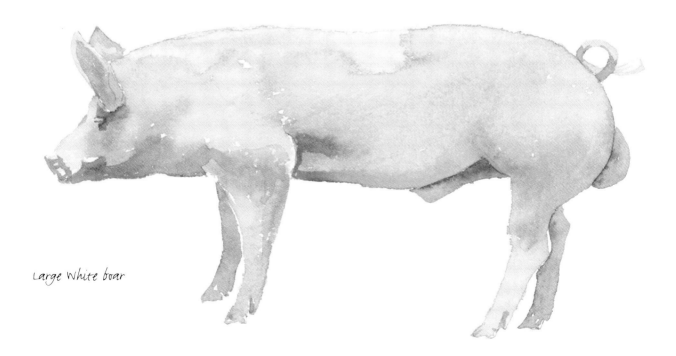

Large White boar

Large White piglet

Large White gilt

ORIGIN	TYPE	SIZE	EARS	CHARACTER
UK	Bacon	Large	Prick	Fairly docile but lively

COLOUR

White.

WHAT TO LOOK FOR – POSITIVE POINTS

Moderately long head with slightly dished face. Long slightly arched back with broad well muscled hams.
At least 14 sound and well spaced teats. Skin fine and white.

WHAT TO LOOK OUT FOR – NEGATIVE POINTS

No wrinkles or blue or black spots. No roses or whorls in coat.

Limousin

A black and white beauty from central France

The Limousin or Cul Noir – literally translated as 'black bottom' comes from the Haute Vienne region of the Western Massif Central. It was first recorded in the sixteenth century when records show that the meat was salted and used to provision ships. Traditionally this was a pig that foraged for chestnuts and acorns, was slow to mature and was slaughtered just before Christmas.

These pigs are characterised by their black bottoms and can have one or sometimes even two white rings similar to a Saddleback. A lively and alert pig, the Limousin is quite capable of foraging for a good deal of its food and has low nutrient requirements. The oak and chestnut forests of the Limousin area suit it perfectly and are important in producing the highly prized pork and ham

The average litter is ten piglets and unusually the mothers generally only have 10 teats, but they have good maternal instincts.

DANCING PIGS

A naughty old hog from France
Took a frisky young gilt to a
    dance
But she oinked 'oh you rotter
You trod on my trotter'
And that was the end of their
    prance

Anon

Limousin

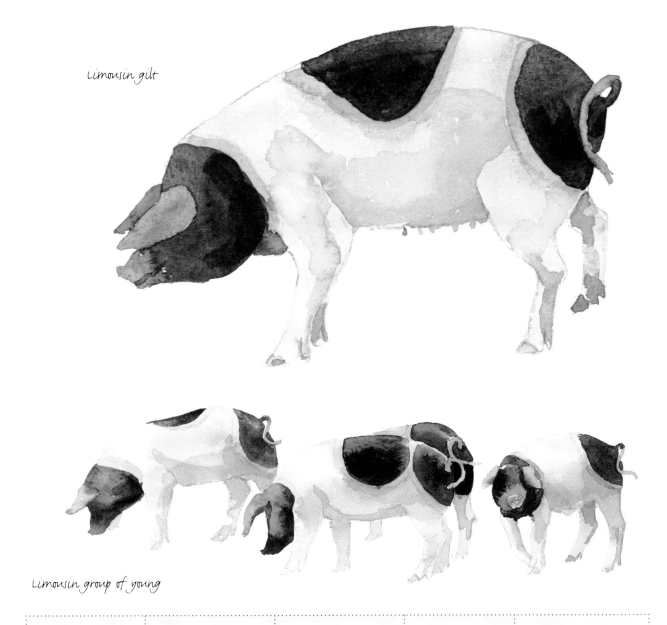

Limousin gilt

Limousin group of young

ORIGIN	TYPE	SIZE	EARS	CHARACTER
France	Pork	Large	Semi-lop	Lively

COLOUR
Black and white, always with black rear.

WHAT TO LOOK FOR – POSITIVE POINTS
Ears pointing forward. Stocky look. Black tail.

WHAT TO LOOK OUT FOR – NEGATIVE POINTS
Less than 10 teats.

Mangalitza

Is it a sheep? No it's a delightful woolly pig

This attractive and unusual pig is all that we have left of the Lincolnshire Curly Coat pig that became extinct in the 1970s. In 1900 LCCs were sent to Hungary to improve the Mangalitza and so some of their blood remains in the current stock. At one time Mangalitzas were common across Germany, Hungary, Romania and Switzerland but they too faced extinction and were brought back from the brink by dedicated breeders in Hungary. They were exported to the USA and Great Britain in 2006 and are now regaining their former popularity.

A lively and friendly creature that would be an excellent choice for first time pig keepers, the Mangalitza is a slow maturing pig. It is very hardy with a good thick coat – but it is also sun tolerant and takes easily to any climate.

DON'T SAY 'PIG' AT SEA

West Indian fishermen who plied the Atlantic never spoke the word 'pig' out loud. They had great respect for pigs because they possessed cloven hooves like the devil. The pig also represented the Great Earth Goddess who controlled the winds. So when at sea, sailors would refer to the animal by such safe nicknames as Curly-Tail and Turf-Rooter. It was believed that mentioning the word 'pig' would result in strong winds, while the killing of a pig on board ship would result in a full-scale storm.

Mangalitza (swallow bellied)

Mangalitza sow with two piglets

The sows produce rather small litters averaging about six stripey piglets. These pigs are increasingly in demand for forestry projects and for producing ham and salami from their well-marbled meat. The fat has a high level of monounsaturated fat that makes it ideally suited for long curing and also has a good balance of omega 3–6 fatty acids. The coat is also used to create fishing flies for fishermen.

Mangalitza sow with piglets

The Dunmow Flitch Trials are held every four years at Great Dunmow in Essex, England. If a married couple can satisfy a jury of six maidens and six bachelors and swear that they have 'never wished themselves un-wed for a year and a day' they are rewarded with a side of bacon or 'flitch'.

ORIGIN	TYPE	SIZE	EARS	CHARACTER
Europe	Lard	Medium	Semi-lop. Prick in red	Friendly

COLOUR

Blonde, red and swallow bellied (black with pale underbelly).

WHAT TO LOOK FOR – POSITIVE POINTS

Dense long hair, curly in winter but shorter and straighter in summer.
'Well-man spot' a bright area of skin 3 to 5cm (1–2in) in diameter, usually found near the ear.

WHAT TO LOOK OUT FOR – NEGATIVE POINTS

Less than 10 teats. Skin should not be spotted. Hooves should not be grey or yellow.

Meishan

A wrinkly producer of very large litters

Meishan pigs come from the mild regions of Central China and were first imported into the USA in 1989. This is a slow growing breed that can become excessively fat but has a reputation for producing meat with an excellent flavour. The outstanding feature of a Meishan is its extraordinary wrinkled skin, particularly around the face.

This is a pig that will do best free ranging where it can forage for itself – in its native China it would have lived near lakes and eaten water plants growing round the shore. The Meishan is one of the most prolific pigs in the world, producing litters of up to 16 piglets. It reaches puberty at the early age of two and a half to three months of age and the sows frequently have two litters a year. The sow's belly can become so low in pregnancy that it is almost dragging along the ground. It also has a reputation for being resistant to disease.

The Meishan is a truly docile pig, some might say lazy, which makes it an ideal and interesting breed for the small holder.

FAT PIG

Ton Pig was a domestic hog from North Eastern China that grew to weigh an astonishing 900kg (1,984lb). Originally the runt of the litter, Ton was lovingly cared for and fed by his owner Xu Changjin. He slept in Xu's bedroom, eventually moving to the living room as he grew, until he became too large even for that. When, sadly, Ton died from lack of mobility due to his extreme obesity on 4 February 2004, Xu is said to have cried for a week. He donated Ton's body to the Liaoning Agriculture Museum in Shenyang and it took 16 people to load him onto the truck for the journey to the taxidermist.

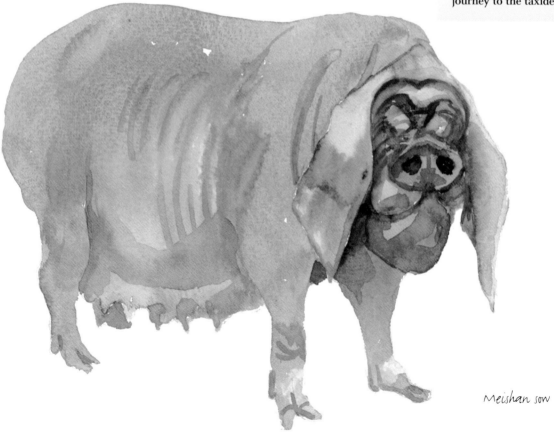

Meishan sow

Meishan piglets

THE YEAR OF THE PIG

The pig is one of the animals of the Chinese zodiac. Believers in Chinese astrology believe that each animal in the cycle has certain personality traits and the pig is associated with fertility and virility. A mother who bears children in the Year of the Pig is considered very fortunate – for they will grow to be happy and honest.

Meishan sow

ORIGIN	TYPE	SIZE	EARS	CHARACTER
China	Pork	Small	Lop	Docile

COLOUR

Black or grey occasionally with white or pinkish legs and belly.

WHAT TO LOOK FOR – POSITIVE POINTS

Wrinkles all over body particularly on the face.

WHAT TO LOOK OUT FOR – NEGATIVE POINTS

Lack of distinctive wrinkles. Belly too low to ground.

Middle White

Beauty is in the eye of the beholder...

The origin of this pig is credited to one Joseph Tuley who exhibited a collection of his pigs at Keighley Agricultural Show in 1852. The judges decided they were neither Large Whites nor Small Whites (now extinct) and so a third class was established. They were in fact crossings of the Large and Small and gained their snub noses from the Small White that had Siamese and Chinese pigs in its make up. Thanks to its strange looks it is sometimes called the 'Bat Pig'.

Very popular before the Second World War thanks to its smaller sized joints, the Middle White suffered after the war when food rationing created more of a demand for the larger bacon pigs.

This is an ideal beginners' pig, not being too large and with a calm and docile temperament – its upturned nose also means that it tends to do less rooting than some other breeds. It is however, not as hardy as some and needs good housing during the winter. The Middle White is an early to mature breed, much favoured by butchers and has now been exported worldwide. It is particularly popular in Japan where it is known as the Middle York.

A LAIRD AND HIS PIG

Lord Gardenstone, an eighteenth-century Scottish judge, had two favourite tastes: he indulged in the love of pigs and snuff. He took a young pig as a pet, and it became quite tame, and followed him about like a dog. At first the animal shared his bed; but when it became unfit for such companionship, he still allowed it to sleep in his room, on a comfortable couch formed of his own clothes.

The Book of Scottish Anecdote,
1883

Middle White boar

Q: What do you give a pig that's sick?
A: Oinkment.

Q: Where do pigs go on holiday?
A: The tropigs.

Q: What sort of dog would a pig own?
A: A pigenese.

Middle White with day-old piglets

Middle White gilt in-pig

ORIGIN	TYPE	SIZE	EARS	CHARACTER
UK	Pork	Medium	Prick	Prick

COLOUR
White.

WHAT TO LOOK FOR – POSITIVE POINTS
Moderately short dished face. Fairly large ears inclined forward. Straight short legs standing well up on its toes.

WHAT TO LOOK OUT FOR – NEGATIVE POINTS
There should be no spots on the skin. Extra toes or a twisted jaw.

Mora Romagnola

A rare pig with Latin looks

Originally found on the Northern side of the Italian Appennines, the Mora Romagnola is an old breed that was once very numerous but is now classed as rare. However, the WWF Italy, in collaboration with Turin University, has put in place a recovery plan and numbers are increasing once more. Initially there were three varieties, the Faentina that was red, the Riminese, a darker red and the Forilvese, similar to the dark Mora of today. They were also crossed with Large Whites and the resulting hybrid was known as Fumati or Smokey.

Unusually, in this breed the sow is larger than the boar. Being a fairly primitive breed they tend to put on fat and although slow to mature are vigorous grazers, feasting on whatever local food is available. This is most likely to be the chestnuts and acorns that produce a flavoursome meat now popular with more sophisticated tastes.

THE ORIGIN OF THE BARBECUE

The word 'barbecue' is thought to have come to Europe via returning Spanish explorers. They had met Arawak Indians who erected wooden frames over fires to dry meat. In the Arawak Taino language this process was known as 'barabicu' which was Europeanised as 'barbacòa'. The word 'barbe à queue' is different, a reference to the fact you can eat a pig 'from the beard to the tail'.

Mora Romagnola boar

You gotta have swine to show you where the truffles are

Edward Albee

The sows give birth out in the field, finding a safe corner for themselves, but rarely have large litters. The piglets are dark golden when born but become darker with age.

Mora Romagnola gilt

Mora Romagnola
sow with piglet

ORIGIN	TYPE	SIZE	EARS	CHARACTER
Italy	Pork	Medium	Lop	Vigorous

COLOUR
Black to dark grey with pink abdomen.

WHAT TO LOOK FOR – POSITIVE POINTS
Long nose with ears falling gently forward. Almond shaped eyes.

WHAT TO LOOK OUT FOR – NEGATIVE POINTS
Over large head. Too fat.

Mulefoot

A very rare breed with a mono trotter

This strange name for a pig derives from the fact that the Mulefoot has a syndactyl hoof, a term meaning 'digits fused together', that makes the trotter resemble that of a mule. This is a breed unique to the USA and is sadly now critically rare, although certain dedicated breeders are trying to raise awareness of its state and popularity is slowly increasing.

It is thought that the original Mulefoots arrived on the Gulf coast with Spanish invaders though this is far from certain. Whatever its origins, by the early 1900s the Mulefoot was a popular backyard pig, valued for its ease of fattening and production of good quality ham. It was particularly numerous in the Corn Belt states and along the banks or islands of the Mississippi and Missouri rivers – perhaps because its uncloven hoof was well adapted to boggy land.

Mulefoot sows make calm and good mothers though litter sizes are mostly small, indeed this is a calm and friendly breed altogether and ideal for the first timer or smallholder.

THE HEAVENLY SOW

Nut, the Egyptian sky goddess and goddess of the night, whose image was painted underneath the lid of coffins, was often depicted as a heavenly sow, eternal mother of the night stars, who were identified as thousands of piglets.

Mulefoot piglets in feed bowl

Mulefoot gilt

MULEFOOT

Whether you ignore a pig or worship that
pig from afar, to the pig it's all the same.

spanish proverb

Mulefoots

ORIGIN	TYPE	SIZE	EARS	CHARACTER
USA	Lard	Medium	Semi lop to prick	Characterful

COLOUR
Black with occasional white points.

WHAT TO LOOK FOR – POSITIVE POINTS
Solid hoof uncloven. Wattles are allowed. Soft silky coat.

WHAT TO LOOK OUT FOR – NEGATIVE POINTS
Too much white is undesirable. Any sign of a cloven hoof unacceptable.

Ossabaw

An unusual feral hog left behind by the Spanish in America

Ossabaw is an island off the coast of Georgia and the Ossabaw Island Hog supposedly arrived there with Spanish explorers in 1500. They brought pigs with them and deposited a few on islands close to where they landed as a kind of larder. The pigs adapted to their island home and since the food available was not always nourishing or readily obtainable they gradually became smaller – a process known as 'insular dwarfism'. They quickly put on fat, a trait that enabled them to survive when times were lean. Their ability to store fat becomes apparent if they have too little exercise when they will soon become obese. They also developed a salt tolerance that helped during periods of drought.

... there warn't anybody at the church, except maybe a hog or two, for there warn't any lock on the door, and hogs likes a puncheon floor in summer-time because it's cool. If you notice, most folks don't go to church only when they've got to; but a hog is different.

Mark Twain,
The Adventures of Huckleberry Finn

Ossabaw sow

The Ossabaw is a small, intelligent and friendly pig and although it is an excellent forager, grows rather slowly in comparison to other breeds, thereby producing top quality flavourful pork. In captivity they will gradually become larger. They are easy to train and can live for up to 25 years.

These pigs also have a form of low-grade non-insulin dependent diabetes that makes them excellent animals for medical research. Although most of the population is still on Ossabaw Island where it is kept under quarantine, there are a few to be found on the mainland and if available would be an excellent exotic breed for the smallholder.

Ossabaw gilt

Ossabaw piglet

ORIGIN	TYPE	SIZE	EARS	CHARACTER
USA	Feral	Small	Upright	Friendly

COLOUR
Black, black and white spotted or red and tan.

WHAT TO LOOK FOR – POSITIVE POINTS
Long snout. Light rear quarters. Heavily bristled.

WHAT TO LOOK OUT FOR – NEGATIVE POINTS
Excessive obesity.

Oxford Sandy and Black

Charming all-round pig with unusual colouring

Also known delightfully as the 'plum pudding' pig, the Oxford Sandy and Black or Oxford Forest pig comes in a range of colours from pale sand to a rich dark chestnut but always with black blotches (not spots). This is an attractive pig with semi-lop to full lop ears and slightly dished muzzle. It is a natural forager and browser that produces good quality meat and tends not to run to fat.

An old breed, known since the 18th century, it may have developed from crosses of Old Berkshire with Tamworth, with a bit of Gloucester Old Spot thrown in, although its true ancestry is unrecorded. Occasionally a pig with prick ears showing Tamworth heritage is born, but this is not acceptable in the breed standard and should not be bred from. The popularity of the OSB as gone up and down in the past and it has faced extinction on more than one occasion but currently its survival seems assured and it has had its own breed society since 1985.

Known for its gentle temperament and good mothering abilities, this would be a good choice for the first time pig keeper since it is a hardy beast and doesn't suffer from sunburn thanks to its coloured coat. It is a traditional cottager's pig that finishes quicker than some breeds and thrives under most management systems.

A PIG ON A CASTLE

The Pig and Tinderbox is an old name for the Elephant and Castle public house: an allusion to its symbol of an elephant surmounted by an erection resembling a castle.

Oxford Sandy and Black

Barber, barber, shave a pig;
How many hairs will make a wig?
'Four and twenty, that's enough.'
Give the poor barber a pinch of snuff.

Anon

Oxford Sandy and Black piglet

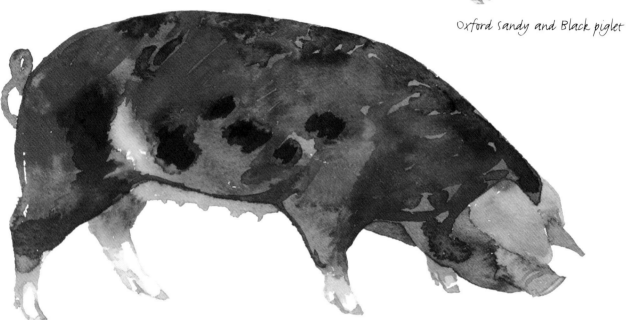

Oxford Sandy and Black sow

ORIGIN	TYPE	SIZE	EARS	CHARACTER
UK	Pork	Large	Semi-lop to full lop	Docile

COLOUR

Ground colour sandy, ranging from pale sand to rust. Markings black in random blotches rather than small spots with sandy the predominant colour. Pale feet, blaze and tassel are characteristic.

WHAT TO LOOK FOR – POSITIVE POINTS

Long deep body with broad hindquarters and rather finer forequarters. Moderately long head with slightly dished muzzle.

WHAT TO LOOK OUT FOR – NEGATIVE POINTS

Short or very dished face a defect. Erect ears unacceptable.

Pietrain

A rare but docile pig ideally suited to the smallholder

Originally bred in the village of Pietrain in the Walloon Municipality of Jodoigne, in Belgium, this breed was developed by crossing local swine with Gloucester Old Spots imported from Britain. The original GOS sow was called Esperance de la Sarte and nearly all Pietrain pigs can trace their ancestry back to her.

The crossing was very successful and progeny were exported to many other European countries, in particular large pig producing countries such as Germany and Spain, where they was used as terminal sires to improve other breeds. Pietrains arrived in the UK in 1964.

THE BOAR'S HEAD

The Old English custom of serving a boar's head as a Christmas dish may derive from Scandanavian mythology. Freyr, the god of peace and plenty, rode Gullinbursti the golden boar; his festival was held at Uletide (winter solstice) when a boar was sacrificed to his honour.

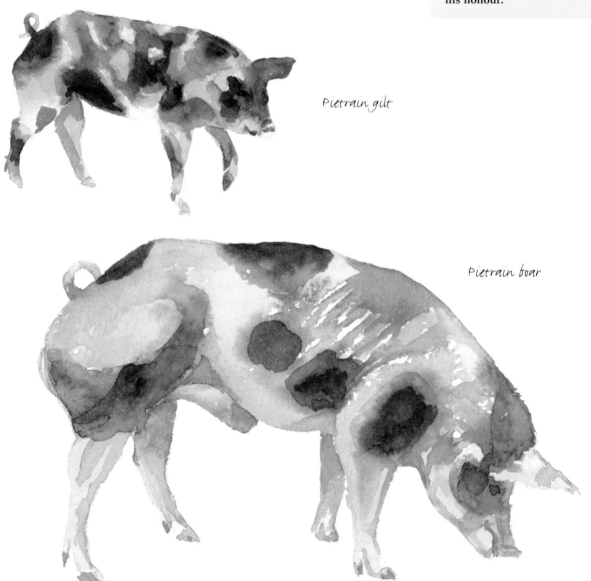

Pietrain gilt

Pietrain boar

The pigs are mainly white with black splotches surrounded by grey rings where the white hair grows through pigmented skin – making them effectively piebald. This is a muscular pig with a very high yield of lean meat. It was unfortunately associated with the gene for Porcine Stress Syndrome though this has mainly been bred out.

The sows are exceptionally docile and easy to handle and make good milky mothers of large litters.

Pietrain gilt

Pietrain gilt

ORIGIN	TYPE	SIZE	EARS	CHARACTER
Belgium	Pork	Medium	Semi-lop	Docile

COLOUR

White with black patches.

WHAT TO LOOK FOR – POSITIVE POINTS

Relatively small head with broad straight snout. Horizontal ears neither lop nor prick, pointing forward. Very well muscled. Furrow along vertebral column accentuating squareness.

WHAT TO LOOK OUT FOR – NEGATIVE POINTS

Ears that are either prick or lop are incorrect. Lack of muscle. Predominance of black skin.

Poland China

A large, gentle pig that produces superb meat

Since this breed came neither from Poland or China the name is somewhat of a mystery. It is thought to come from the fact that in the 1860s a Polish farmer in the US developed an exceptionally large breed that had been crossed from local pigs, Chinese pigs and Berkshires.

Originally the Poland China was bred for lard but today the aim is pork and their conformation has been adjusted to produce a less fatty pig. They are known for their quick growth and top quality marbled meat.

BIG BILLY

In 1933 a Poland China hog named Big Billy weighed in at 1157kg (2552lb). He was 1.5m (5ft) tall at the shoulder and 2.74m (9ft long). He was so fat that his stomach dragged on the ground. By 1998 when the breed had been reduced in size the record breaker named Big Blue only weighed 480kg (1060lb) – about the size of a small car.

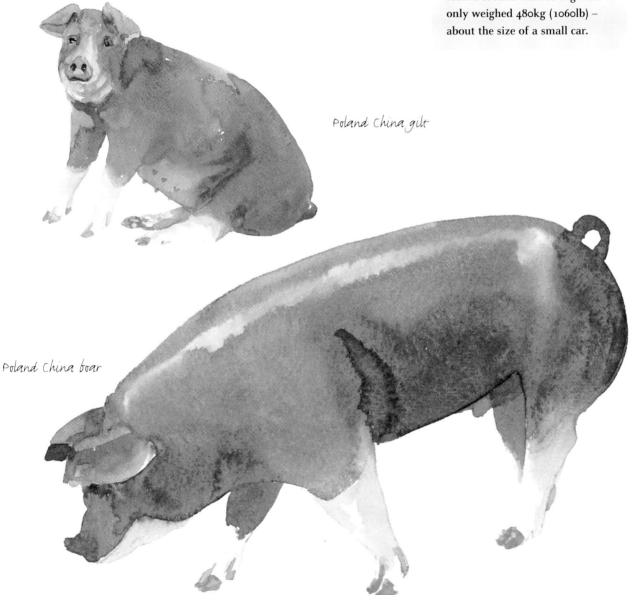

Poland China gilt

Poland China boar

The Poland China does not do well in a commercial indoor unit but thrives outside and its docile and friendly nature makes it an ideal smallholder's pig. The sows are good mothers, producing large litters and plenty of milk. The Poland China was one of the first pig breeds developed in the United States and was instrumental in the development of the Spotted breed (page 130).

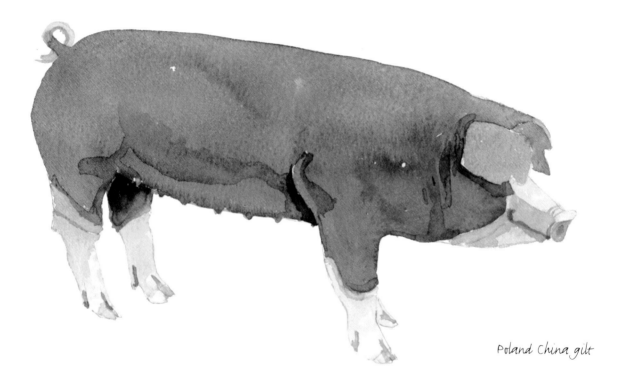

Poland China gilt

ORIGIN	TYPE	SIZE	EARS	CHARACTER
USA	Pork	Medium to large	Lop	Gentle

COLOUR
Black with six white points (face, feet and switch).

WHAT TO LOOK FOR – POSITIVE POINTS
Ears must be down. No evidence of a belt visible.

WHAT TO LOOK OUT FOR – NEGATIVE POINTS
Should not have red or sandy hair. A hog may not have more than one solid black leg.

Red Wattle

A friendly red pig with a tasselled accessory

The unusual but not unique feature of the Red Wattle is its wattle. This is a fleshy skin covered tassel that hangs each side of the neck similar to that of a goat. There is apparently no function to this appendage but it is a single gene characteristic and is passed on to any progeny when crossed.

No one is sure where the original Red Wattles came from but a number were living a semi-wild existence in Texan woodland in the early 1970s. A farmer there, a Mr H. C. Wengler, crossed them with a Duroc and then put the offspring back to the original sow to create what became known as Wengler Red Wattle Hogs. In the 1980s another line known as Timberline was crossed with a Wengler Red Wattle to create the Endow Farm Wattle Hogs. In 2001 the American Livestock Breeds Conservancy set up the Red Wattle Hog Association to centralise registry.

A PIG AND HIS HOBBY

There was a young pig whose
   delight
Was to follow the moths in their
   flight.
He entrapped them in nets,
   Then admired his pets
As they danced on the ceiling at
   night.

          Anon

Don't give that which is holy to the dogs, neither throw your pearls before the swine, lest perhaps they trample them under their feet, and turn and tear you to pieces.

Matthew 7:6

Red Wattle gilt

Red Wattles are adaptable pigs that will be happy in any climate. They are hardy and good foragers that put on weight fast. Proponents of the breed insist that there is no meat tastier than that produced by these pigs. The sows have large litters and milk to match and their placid temperament makes this as ideal breed for the novice pig keeper.

Red Wattle sow with piglets

ORIGIN	TYPE	SIZE	EARS	CHARACTER
USA	Bacon	Large	Erect, tipped or lop	Gentle

COLOUR

Red but ranges from almost yellow to almost black.

WHAT TO LOOK FOR – POSITIVE POINTS

Sturdy straight legs set well apart. Firm slightly arched back. Two well formed wattles.

WHAT TO LOOK OUT FOR – NEGATIVE POINTS

Lack of wattles.

Spotted

A pig that prefers the outside life

The Spotted or Spots breed was at one time known as the Spotted Poland China but became known simply as the former in 1960. The breed was created in Indiana from a crossing of local stock with Poland China hogs and Gloucester Old Spots. The pig that emerged was a feed-efficient animal that gained weight fast and passed on its good qualities to its offspring.

Spotted swine do not take to commercial indoor breeding but flourish in an outdoor environment being hardy big-boned creatures. Ideally the colouring should be equally divided between black and white with grey edges to the blotches caused by white hair growing from pigmented skin. To be registered they must have at least 20 per cent and no more than 80 per cent white on their bodies.

The ball used in American football used to be called the 'pigskin' after the material used in its manufacture – but today most are made of cowhide.

THE GREAT PIG WAR

In 1859, an unnamed British-owned pig wandered into a man called Lyman Cutlar's potato patch on San Juan Island off the coast of Vancouver. He shot the pig, setting off a military stand-off that became known as the Great Pig War. Also called the Pig Episode and the Pig and Potato War, this was a confrontation between American and British authorities over the boundary between the USA and British North America. When British authorities threatened to arrest Cutlar, American citizens drew up a petition requesting US military protection. The episode began a 13-year stand-off between the US Army and the British Royal Navy. The pig was the only casualty.

spotted boar

spotted gilt

ORIGIN	TYPE	SIZE	EARS	CHARACTER
USA	Meat	Medium to large	Semi-lop	Gentle

COLOUR

Half and half black and white spotty blotches with grey edging.

WHAT TO LOOK FOR – POSITIVE POINTS

Long back. Muscular build. Forward inclined ears. Dished face with medium to short snout.

WHAT TO LOOK OUT FOR – NEGATIVE POINTS

More than 80 per cent black or white colouring. Brown or sandy spots. Completely black face.

Swabian Hall

A producer of highly prized pork

This handsome German pig comes from Schwabisch Hall, Baden Wurttemberg, in southern Germany. In 1930 King Wilhelm I of Wurttemberg imported Chinese Meishan pigs to cross with the local Landrace in order to increase their fat content. As modern tastes prefer a leaner meat with less fat, the numbers of this breed have subsequently declined.

The Swabian Hall's survival is thanks to the dark meat and strong flavour, and its pork is highly prized by gourmets and the name is protected. Only pigs from the Schwabisch Hall and nearby regions can be sold under that name and all producers belong to a breed association that regularly inspects the quality of the food given to the animals. Its diet should be mainly cereals, peas and beans and should not include any drugs or antibiotics or genetically-modified feed.

This is a highly fertile breed, long-lived and hardy, the sows making excellent mothers with abundant milk.

In Iowa, USA a pig breeder called Carl Blake crossed a Meishan with a Russian Wild Boar to try and create the original breed. He named it the Iowa Swabian Hall and reckons it rivals the German version in all but colour – his are totally black.

SYMBOL OF FERTILITY

The Teutons, one of Germany's ancient peoples, sacrificed a pig – their most valuable animal – to the gods to ensure good luck. To the Teutons, the pig was a symbol of fertility and wealth and would help to ensure the success of a pregnancy.

Swabian Hall sow

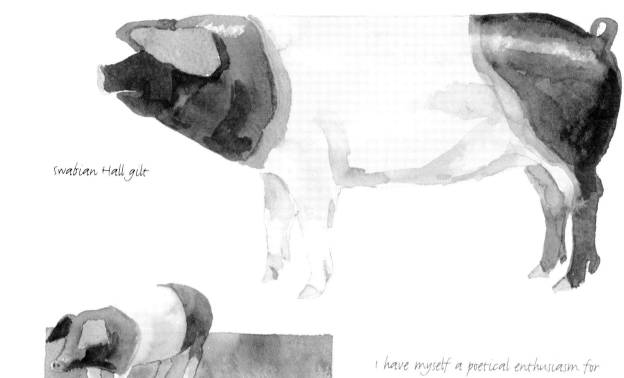

swabian Hall gilt

I have myself a poetical enthusiasm for
pigs, and the paradise of my fancy is one
where pigs have wings. But it is only men,
especially wise men, who discuss whether
pigs can fly; we have no particular proof
that pigs ever discuss it.

G K Chesterton

swabian Hall weaners

ORIGIN	TYPE	SIZE	EARS	CHARACTER
Germany	Pork	Large	Pork	Passive

COLOUR

White belt with black head and rear, white tips to nose and tail.

WHAT TO LOOK FOR – POSITIVE POINTS

Broad white belt that includes front and may also include back legs.

WHAT TO LOOK OUT FOR – NEGATIVE POINTS

Prick ears or those that obscure vision.

Tamworth

A ginger character with an extra long snout

The Tamworth is one of the oldest breeds of pig, descended from native European pigs that had developed from Wild Boar. Named after the Staffordshire town of its origin it was one of the most popular breeds by the middle of the nineteenth century and was exported widely, reaching North America by about 1870.

Originally the Tamworth was bred for bacon – the length and depth of its sides being favoured by bacon curers, but this is an adaptable breed and is also suited to pork production. The Wild Boar ancestry shows in the Tamworth's hardiness and foraging skills. It is also apparent in the pig's ability to run like the wind: the Tamworth is an expert escaper so will need extra strong fencing to contain it. Woodland would be its natural surroundings and the more space available the happier a Tamworth will be.

FIGHTING AN INVADER

In Hampshire's New Forest, pigs have been used to try and rid the forest of an invasive foreign shrub called Gaulthieria. It is native to North America and much enjoyed by deer but in the New Forest its root system, which spreads rather like couch grass, is smothering the native plants. Instead of using pesticides which may harm other plants, pigs are used to root up the shrub.

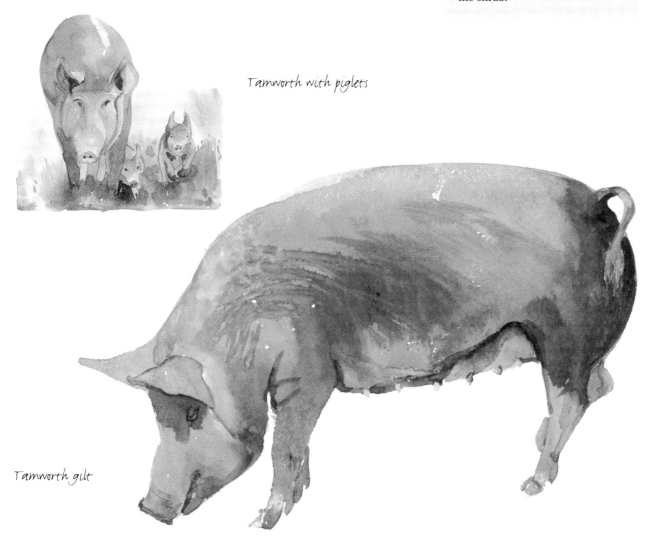

Tamworth with piglets

Tamworth gilt

Tamworth sows make good mothers though they rarely give birth to more than nine piglets and five or six is more usual. This breed is rather slow to mature, a fact that adds to the quality of the meat. Its ginger colouring means that sunburn is not a problem.

Tamworth gilt

THE TAMWORTH TWO

In January 1998 two young Tamworths escaped from an abattoir near Malmesbury in Wiltshire. The brother and sister pigs squeezed under a fence, swam across the River Avon and sought sanctuary in a wood where they remained for over a week. Various methods were used to capture them but they evaded them all. The two became known as Butch Cassidy and the Sundance Pig and their story caught the world's imagination. At night they visited local gardens to search for food and were eventually flushed out by two spaniels. Butch was caught but Sundance evaded capture for longer and had to be darted with a tranquilliser. So popular had they become that they were bought by the *Daily Mail* newspaper and sent to an animal sanctuary where they lived out their lives in luxury.

ORIGIN	TYPE	SIZE	EARS	CHARACTER
UK	Bacon and pork	Large	Prick	Intelligent

COLOUR

Ginger, ranging from golden to dark red.

WHAT TO LOOK FOR – POSITIVE POINTS

Straight face and snout. Long narrow body with deep sides, long neck and legs.

WHAT TO LOOK OUT FOR – NEGATIVE POINTS

Dark spots in the hair or curly coats not acceptable. Short or dish nose unacceptable.

Vietnamese Potbelly

A characterful creature with the largest of bellies

In its native Vietnam the potbelly has the short name 'Í' and in rural regions many families would keep a Potbelly, letting it feed on household scraps and what it could forage for itself.

Vietnamese Potbellies were imported into the United States via Canada in the 1960s as zoo animals. It wasn't until 1985 that they became available as pets. Enormous sums were paid by people beguiled by the frankly ugly faced creature whose main attraction was its small size. Even so potbellies can reach 90kg (200lb) in weight when they reach maturity at around five years of age although the average is 68kg (150lb).

Both sexes reach fertile maturity at a young age and the sows have large litters though piglets grow slowly. If the pigs are kept as pets they are usually neutered or spayed as the sows can become very moody when in season. Their diets must be carefully monitored as they can easily become overweight. Unusually in swine these pigs do not have curly tails (if they do it shows evidence of cross-breeding) and their bristles are exceptionally coarse.

LULU THE LIFESAVER

In October 1998, Jo Ann Altsman collapsed at home with a heart attack. Luckily she was not alone. Her pet Potbellied pig, Lulu, quickly realised that something was wrong and went for help. She managed to break out of the yard and waited in the middle of the road for a passing car. When a car stopped, LuLu led the driver into the house where he called out, 'Lady, your pig's in distress'. Jo Ann replied, 'I'm in distress, too. Please call an ambulance.'

Vietnamese Potbelly sow

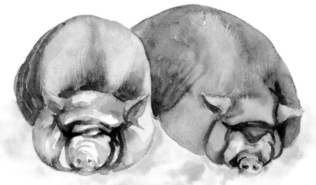

On 24 August 2004, a Vietnamese Pot Bellied pig called Kotetsu became the holder of the Pig High Jump Record. He jumped 70cm (27.5in) at Mokumoku Tedsukri Farm, in Japan's Mie Province.

Vietnamese Potbelly sows

Vietnamese Potbelly gilt

ORIGIN	TYPE	SIZE	EARS	CHARACTER
Vietnam	Pet	Small	Prick	Temperamental

COLOUR
Black but also found in white, grey or rust.

WHAT TO LOOK FOR – POSITIVE POINTS
Sway back normal. Prickly coarse hair.

WHAT TO LOOK OUT FOR – NEGATIVE POINTS
Straight tail. Stomach dragging on the ground.

Welsh

It's good if it all goes pear-shaped

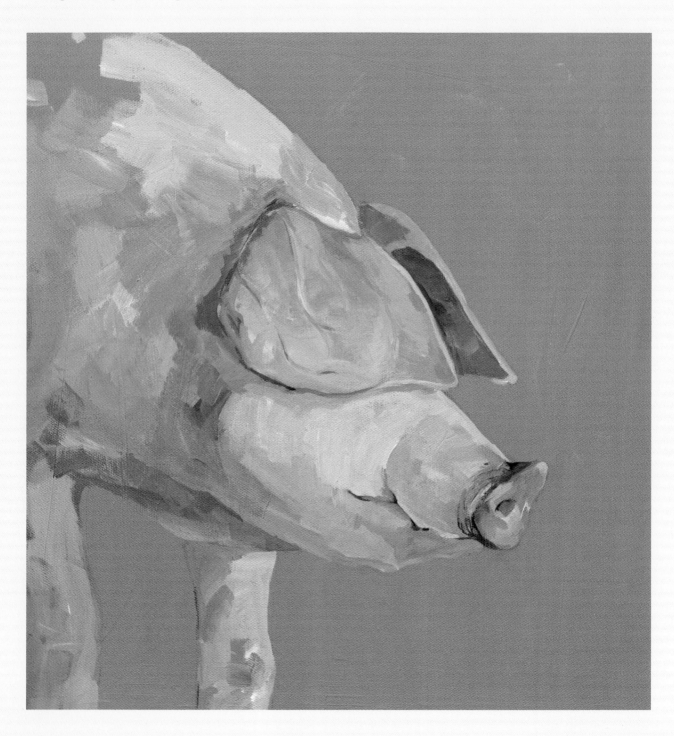

There are records of lop-eared Welsh pigs dating from the 1870s when they were traded into Cheshire to be fattened on dairy by-products. The Old Glamorgan Pig Society was established in 1918 along with a Welsh Pig Society that covered pigs bred in Carmarthen, Pembroke and Cardigan. In 1922 the two Societies amalgamated. After the Second World War Welsh pigs increased in popularity until they were rated number three in the country behind Large Whites and Landrace.

The Welsh is more of a yellow white, rather than pink white, pig with a long body. Ideally it should look pear-shaped when seen from either above or side on with good strong hams. This is a hardy breed that is very docile and easy to manage and would make an ideal novice or smallholder's pig. Sunburn is a problem as with all light skinned animals and so plenty of shade must be available. The sows make excellent mothers of large litters.

Sadly the trend towards leaner pork and bacon has caused a fall in popularity of this breed and it is now on the Rare Breeds List.

THE GODDESS OF THE UNDERWORLD

The Celtic goddess of the Underworld Cerridwen is often symbolised by a white sow, which represents both her fecundity and fertility and her strength as a mother. She was renowned for her magical cauldron. It contained a potion that granted knowledge and inspiration to all who ate from it. The potion had to be brewed for a year and a day to reach its full potency.

Welsh gilt

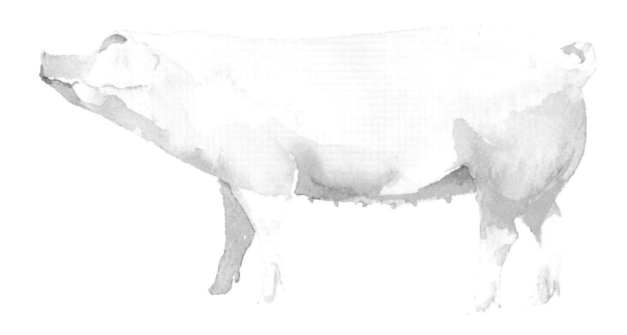

WELSH

To market, to market to buy a fat pig;
Home again, home again, jiggety-jig.
To market, to market, to buy a fat hog;
Home again, home again, jiggety-jog.

To market, to market, to buy a plum cake;
Home again, home again, market is late.
To market, to market, to buy a plum bun;
Home again, home again, market is done.

To market, to market, a gallop a trot,
To buy some meat to put in the pot;
Three pence a quarter, a groat a side,
If it hadn't been killed it must have died
Old nursery rhyme

Welsh sow with piglets

ORIGIN	TYPE	SIZE	EARS	CHARACTER
UK	Pork and bacon	Medium	Lop	Easy to keep

COLOUR
White.

WHAT TO LOOK FOR – POSITIVE POINTS
Straight snout. Level back with straight underline. Lop ears just meeting at the tip of the snout.

WHAT TO LOOK OUT FOR – NEGATIVE POINTS
Blue spots undesirable. No roses in coat.

Wild Boar

A ferocious producer of mouth watering pork

Wild Boars were once found in one form or another all over the world. They died out in Britain around the end of the seventeenth century but thanks to escapes from modern day boar farms, today are found in living wild in a number of areas in Britain. They are still very common on mainland Europe and every year around a million are shot in Germany alone, both for sport and to control the numbers living wild as they can be a severe nuisance to farmers.

In Britain the Wild Boar is classed as a dangerous animal and a licence must be acquired from the local authority in order to breed them. They require extra strong fencing that must be buried at least 30cm (1ft) in the ground. Ideally Wild Boar should be kept in woodland in groups of six to ten sows to one boar. They can live outside all year as they are exceptionally hardy but are nocturnal.

Sows will happily farrow outside, producing small litters of only two or three boarlets for the first time mother rising to six to eight for an older sow. The boarlets are born stripey but gradually lose their stripes as they grow to develop a thick, coarse bristly coat.

THE BLUE BOAR

The cognisance was a distinguishing mark worn by an armed knight, usually on his helmet. The blue boar was the cognisance of Richard III, king of England from 1483 to 1485. In Leicester there is a lane in the parish of St Nicholas, called Blue Boar Lane, where Richard slept the night before the battle of Bosworth Field.

Wild Boar sow

The fact that they are slow to mature helps to produce excellent quality pork. However, their wild characteristics and aggressive nature also renders them totally unsuitable subjects for the novice pig keeper.

Guillaume, Comte de la Marck (died 1485) was known as The Wild Boar of the Ardennes because he as was fierce as the wild boar, which he delighted to hunt.

Brewer's Dictionary of Phrase & Fable

Wild Boar boarlets

Wild Boar boarlet

ORIGIN	TYPE	SIZE	EARS	CHARACTER
Worldwide	Pork	Medium	Prick	Wild and fierce

COLOUR
Greyish black with rusty tints.

WHAT TO LOOK FOR – POSITIVE POINTS
Long thin snout. Ridge of bristly hair along back.

WHAT TO LOOK OUT FOR – NEGATIVE POINTS
Any sign of crossing with domestic pigs. Lack of bristles.

Time for slaughter

*Unless your bacon you would mar, kill not
your pig without the 'R'*

*A reference to the winter months of September to
April when meat is less like to go off in the heat –
commonly also the safe months for eating oysters.*

At around six months of age your weaners will have become porkers and are ready for slaughter. It might seem that if you keep them longer they will be bigger and produce more meat, but in fact after this age they will simply put on fat. You can keep them longer if you decide you want to produce baconers.

Find as small an abattoir as you can where you feel happy that your animals will be treated kindly and book them in. This should be done well in advance as many abattoirs have specific days for pigs. Check that you have room in your freezer and that any customers are also ready to receive their pork fresh (rules apply to selling fresh meat: see www.food.gov.uk).

The last thing you want, probably early in the morning, when the dreaded day arrives is to have a problem loading your pigs onto the trailer – if possible practise this in advance and if practicable, on their last night keep them in an enclosed space that you can back the trailer up to. Make sure you have attached a metal ear tag or slapmark both shoulders with your herd number and fill in the food chain information form that you can get from the abattoir. You will also require a completed movement licence form (AML 2) (see page 15).

Saying goodbye to your pigs is never easy but keep hold of the thought that you have given your pigs a happy life. Remember too that if everyone gave up eating pork, pigs would very soon become extinct.

What happens at the abattoir

You should tell the slaughterman if you want the head or 'pluck' which includes the liver and kidneys along with other bits of intestines or the abattoir may retain them.

The pigs enter the abattoir one by one and are checked by a veterinary inspector to make sure they are fit and well (they can be rejected if they are too dirty). They are then stunned by a powerful electric current applied to either side of the head which renders the pig unconscious and it is hung up by one back leg. The jugular vein is cut and this is what actually kills the pig.

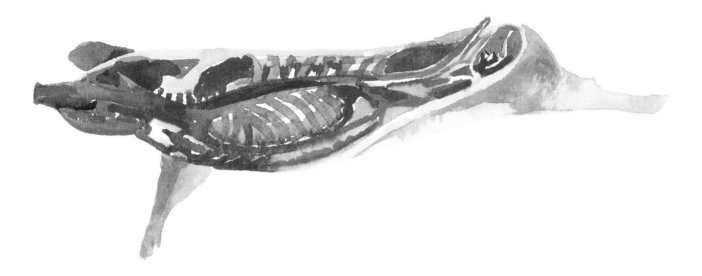

Strict time limits must be adhered to between the stunning and the bleeding. The entire pig is then dipped in boiling water to loosen the bristles which are then removed by a dehairing machine. The carcass is then split and disembowelled and re-inspected by the veterinarian. It is then washed and chilled and hung in a cold room for collection.

Some abattoirs will also butcher the carcass for you, or you can do it yourself. Ideally you should find a professional butcher to do it for you and he will need to know what cuts of what size you want; whether you want sausages or mince and how you want it packaged.

After you take pigs to slaughter for the first time it is a good idea to talk to the slaughterman or butcher and get their opinion on whether you had your feeding regime correct or if the pigs had put on too much fat.

IN MEMORY OF JOHN HIGGS, DIED 1825

Here lies John Higgs
A famous man for killing pigs
For killing pigs was his delight
Both morning, afternoon and night.
Both heats and cold he did endure
Which no physician could e'er cure
His knife is laid, his work is done,
I hope to heaven his soul is gone.

Epitaph in the cemetery of St Mary's Church, Cheltenham

Cuts of me

Any Part of Pig

Any part of piggy
Is quite all right
Ham from Westp
Ham as lean as th
Ham from Virgini
Trotters, sausages,
Crackling crisp for my teeth to grind on
Bacon with or without the rind on
Though humanitarian
I'm not a vegetarian.
I'm neither crank nor prude nor prig
And though it may sound infra dig
Any part of the darling pig
Is perfectly fine with me.

Noel Coward

There are various ways of cutting pork: different countries have different methods and produce different cuts. Before instructing a butcher, decide on the sort of size you want your roasts, whether you prefer to have mince or require the butcher to make sausages, if you like chops or would rather have large joints. The butcher will have his favourite way of butchering but you should always discuss your needs with him.

Categories of meat

In Britain there are four meat categories:

- ❑ **Porker**: 55–62kg (120–136lb) for small joints on the bone, sold as fresh meat
- ❑ **Cutter**: 64–82kg (140–180lb) trimmed of fat and skin
- ❑ **Baconer**: 90–100kg (200–220lb) cured
- ❑ **Heavy hog**: 100–125kg (220–275lb) trimmed of fat and skin

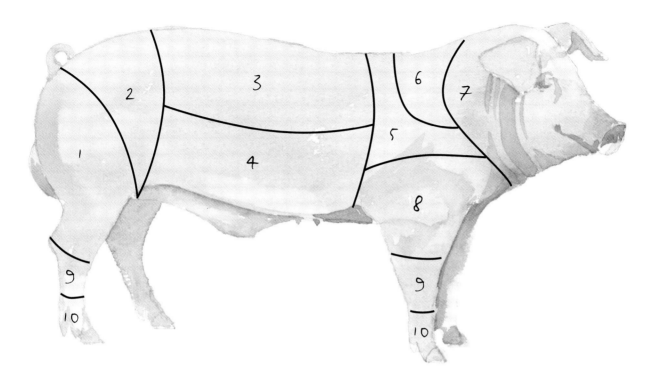

Cuts of pork

1 **Leg** A prime roasting joint that is usually cut into two pieces – the fillet half leg and the knuckle half leg. Whole legs can be boned and rolled by the butcher to the desired size.

2 **Hind loin or chump** Known as chump chops if the bone remains or steaks if it is removed. The fillet or tenderloin is the lean and tender muscle under the backbone.

3 **Loin** The most expensive cut of pork, the loin can be roasted whole but is usually cut into smaller joints or chops.

4 **Belly** Belly pork can be used for stuffing and rolling. Sweetribs come from the rib part of the belly but should not be confused with Spare ribs (6) although they are sometimes also called spareribs.

5 **Blade** A small, cheaper joint that can be roasted or boned and stuffed.

6 **Spare ribs** The cut left on the upper part of the shoulder once the blade has been removed, suitable for roasting, braising or stewing.

7 **Head** Traditionally it is boiled, seasoned, all bones removed and the meat, fat, skin and tongue all chopped, put in a mould and the cooking liquor poured in to make brawn. The cheeks can be used to make sausages with added fat.

8 **Shoulder** A versatile cut, suitable for frying, roasting and slow cooking. The lower part of the shoulder is known as the hand; if the joint includes the jowl it is known as the hand and spring.

9 **Shank** Used in sausages or boiled and potted with belly pork to make rillettes.

10 **Trotter** Usually boiled to make stock.

Cuts of bacon

If you have produced a baconer you don't have to have the entire carcass made into bacon but could keep choice cuts as large roasts. The following are cuts from an entire side that has been wet cured.

1 **Gammon** The hind leg cut square at the top. The most expensive joint, it is suitable for boiling and baking.
2 **Long back** Cut into long rashers.
3 **Oyster** Between long back and top back – cut into very small rashers.
4 **Middle back** A cut that gives a rasher with both back and streaky together, known as an Ayrshire roll.
5 **Back** As a joint it is rolled for boiling, braising or baking. Usually cut into rashers.
6 **Flank** This is where streaky bacon comes from. The rashers always contain some gristle.
7 **Collar** A cheaper joint suitable for boiling. May be cut into smaller joints such as end collar and middle collar.
8 **Fore hock** Front leg usually cut into small joints such as butt, fore slipper or small hock, all suitable for boiling.

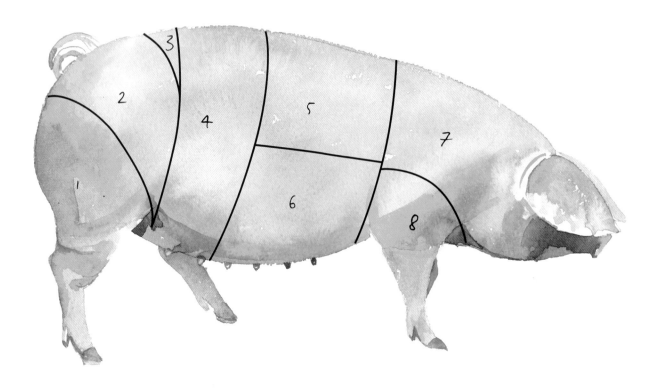

Curing

Bacon is fresh pork that has been preserved or 'cured' with salt so that it will keep longer than if it was fresh. Unsmoked bacon is also known as green bacon. There are two main methods of curing.

Dry curing In this method the meat is rubbed all over with the cure recipe and either started in a fridge or box before being hung in an airy environment to dry – when ready the meat can be eaten raw as with Parma-style ham and salami or cooked as with bacon.

Wet curing Here the meat is immersed in brine. The meat will not keep as long as if it had been dry cured. It will be ready to eat in a much shorter time but will need cooking.

Unfortunately in an era of terrorist bombing it is no longer possible to buy saltpetre (potassium nitrate). It improves preservation and gives the meat its pink look, but is also used in the manufacture of explosives, so cannot be purchased without a special licence. You can mix your own cure – a basic cure will include coarse sea salt (or you can use ordinary kitchen salt, whatever you prefer) sugar and any herbs or spices of your choice, but without the saltpetre your cured bacon will be greyish in colour. Older recipes may call for Prague Powder No 1 or Cure 1 or Prague Powder No 2 or Cure 2, which included sodium nitrate and sodium nitrite, but these are also now unavailable to the public.

However, specialist suppliers of butcher's sundries will supply ready mixed cures that include nitrate in the salt – you simply add sugar and spices. They even supply salami kits with the cure, skin and everything you need – just add meat and fat.

Mix up your cure and keep it in a screw top jar or plastic box and use as needed.

Much smoking kills live men and cures dead swine.
George Dennison Prentice

Making your own bacon

Nothing will be so rewarding or delicious as making your own bacon from your own animals and it is surprisingly simple. The following is a basic recipe for curing bacon but the sky is the limit when it comes to different sugars or spices and you can invent your own recipe. For a really superior taste the bacon should be smoked after it is cured and there are all sorts of ways of doing this, from buying an expensive and sophisticated smoker, to building your own in your garden using an old fridge or barrel (see page 156).

Basic bacon dry cure

You will need:

> 1kg (2lb, 3oz) salt cure
>
> 500g (1lb, 2oz) soft, brown sugar, Demerara or a mix of both
>
> 1 teaspoon of your choice of red or black pepper/nutmeg/powdered clove/juniper berries/cumin or anything you like
>
> (150g (5oz) of this mixture will cure 1kg (2lb, 3oz) of bacon)
>
> A piece of belly pork for streaky or loin for a long back rasher, weighed

1. Rub the cure all over the piece of pork getting in to every nook and cranny. Should you have a piece with a bone, push the cure as far up against the bone as you can. For very thick pieces, pierce the meat randomly with a skewer.
2. Place in a container in the fridge and put a weight on top to speed up moisture loss.
3. After 24 hours drain off any liquid that has appeared (this is also a matter of opinion: some people prefer to leave it) and turn the meat over.
4. Repeat for two more days. Rinse in cold water and dry. Wrap in muslin and hang in a cool dry place for 2 days (or put it back in the fridge if you don't have anywhere suitable). For a very large piece add a day or two, up to a maximum of five days.
5. Test a little bit of the meat in a frying pan and if it is too salty, soak in water for an hour or two.

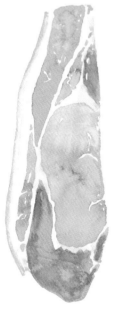

Basic bacon or gammon wet cure

You will need:

> 1kg (2lb, 3oz) coarse sea salt or ordinary kitchen salt
>
> 500g (1lb, 2oz) brown sugar
>
> 1 teaspoon of a spice of your choice such as pepper or cloves
>
> 4.5l (1 gallon) water
>
> A piece of belly pork for streaky or loin for a long back rasher, weighed

1. Warm the water – it does not need to boil – add the salt and stir until dissolved. To check the brine's concentration drop in an un-cooked egg and see if it floats. The egg will sink in plain water but float when the brine is perfect for curing. Add the sugar and spices and dissolve in the water.
2. Submerge your ham and leave for a week for a small ham and 10 days for a large one, turning daily and making sure the entire joint is under water by placing a board or plate on top with a weight on it. After the week or 10 days is up take the ham out and rinse under cold water. Wrap in muslin and hang in a warm place for two to three days. Your ham is now ready to cook.

Air-dried ham

Everyone has heard of Parma ham and many other European countries also produce their own air-dried ham; versions include Bayonne or Serrano. The flavour can be very much a regional attribute depending on the pigs' diet, the quality and dryness of the air and the length of hanging. There is no reason why you should not produce your own with its own distinctive flavour. The main problem with air-drying in Britain is that the air can be damp for months on end, so this is very much a winter project or flies will become an enemy.

Traditionally the whole leg with the bone in would be cured, but for your first attempt try a piece with the bone removed, as there is less risk of the meat going off.

You will need:

Wood or plastic box large enough to take the ham comfortably with a few drainage holes in the bottom,
A weight and some muslin
1 leg of pork carefully weighed
3–5kg (6lb, 8oz to 10lb, 3oz) of the basic bacon dry cure mixture, depending on the size of the pork and its fit in the box
Add whatever flavouring you fancy

1. Place some of the cure mixture over the base of the box – put the leg on the salt in the box (this is where you will discover that the better the leg fits into the box the less salt will be required). Pour in the rest until the leg is completely covered. Put a piece of wood or plastic chopping board on top and place the weight on that.
2. Leave the box in a cool dry room, larder, garage or cellar, being aware that liquor may seep through the drain holes. Cure for not less than three days per kilo and not more than four.
3. Remove the leg from the salt and rinse under cold water, dry and wrap in muslin and tie up with string. Hang it up in as well ventilated and dry a place as you can find. You could make a purpose-built wire netting cage and hang it in your porch, or use a shed with a window open. A larder may not be ideal as they tend to lack air circulation, but this will be a matter of trial and error and yours may be perfect.
4. After four to six months take down your ham and unwrap it. It should be firm to the touch, but not rock hard and you may find that it is covered in mould. This is normal and can be removed by scrubbing with dilute vinegar.
5. Carve the ham as thinly as possible using a very sharp knife, or a slicer if you have one. You may find the ham too salty in which case next time you try, soak the leg in water after the initial salting but before hanging. The fact that it is the salt that is stopping the meat going bad must be taken into account.
6. Serve the ham with something sweet such as melon if you want to counteract the saltiness. You could invent your own accompaniment, such as fig jam or stewed pears in port.

Salami

Salami was developed as a way of preserving meat through the summer months. The salty dry sausages were often smoked as well to add to the flavour and increase their shelf life.

Salami is the ideal way to use up all the left over bits from your carcass. The meat needs to be as lean as possible and then back fat is added. The best way to try is to buy a salami making kit from a butcher's sundries supplier. It will include salt with nitrite added, seasoning of your choice and a starter culture. You will also need some casings known as ox-middles. If mixing the meat by hand use surgical gloves.

The kit will come with its own instructions but the basic method is to chill the meat before starting then cut it into cubes, sprinkle with the seasoning and mince. Add the fat, culture and salt and mix very well before passing it through the mincer again. Fill the ox-middle casing using either a sausage making attachment or by forcing it through a funnel and tie the ends with twine, including a loop to hang it up by.

Put the sausage in a warm environment (15–20°C/59–68°F) for 24 to 36 hours to get the fermentation process going. Then hang it in a dry airy environment, though with some humidity or it will dry out too fast. Leave it for four to six weeks. Any superficial mould can be wiped off using diluted vinegar.

Home smoking

There are plenty of expensive home smokers on the market but with a bit of ingenuity you can build your own. Your main requirement is a 6m (20ft) slope between the fire and where the bacon hangs as it is important for the smoke to be cold. Hot smoke will cook your bacon and this is not what you want.

First, make a covered fireplace – you will need to be able to adjust the draught and it must have an old drain pipe or similar coming out of the back at the top that is connected to the bottom of the smoking chamber far enough away for the smoke to be cold. A barrel is an ideal smoking chamber, but a dustbin or old fridge will also suffice. The salted bacon is hung inside and there must be room for the smoke to escape. (Remember that once you have built your own smoker you can smoke bacon but also salmon and anything else you fancy.).

The fire needs to smoulder for 24 hours or so. Use only hardwoods such as oak and birch. You need are a mix of sawdust and wood shavings, but it is worth experimenting with whatever you can find in your area. The fire will need a fair bit of tending but this is also a matter of practice – have a go and see what flavours you can create!

Sausages

Sausages are known to have been around for nearly 3000 years. One of the earliest mentions of sausages was in Homer's Odyssey, 2700 years ago. They were also a favourite food of the Romans. The word 'sausage' comes from the Latin 'salsus' meaning salted and may well have referred to any cured or preserved meat. Now sausages are found all over the world in one form or another with countries or even cities or counties developing their own special types including Frankfurters or Cumberland sausage.

Around the world sausages also came to have different names – in Australia sausages are known as 'snags' probably from the old English meaning of 'morsel' or 'light meal'. In Britain during the Second World War they became known as 'bangers' as they contained so much water to mask the lack of meat that they tended to explode when fried.

Basic sausage recipe

If you keep your own pigs and have them butchered there will be plenty of meat left over to make sausages and you can develop your own favourite flavour. Getting the sausage meat into the casings is fiddly and takes a bit of practice. You will need casings, obtainable from suppliers of butcher's sundries. You can use a special sausage nozzle, a piping bag or there are fittings for electric mixers available, even specific sausage makers.

A mix of fatty and lean pork is ideal and a small amount either of breadcrumbs, crushed rusks or even oatmeal can be added. The herbs and spices you put in are entirely up to you, but start with salt and pepper and then add whatever you like. Sage is traditional but you could also use ginger, cloves, nutmeg or even curry powder – that is the beauty of a home made sausage, it is entirely your creation. Whatever you choose, write it down so that you can repeat it (or not) again.

You will need:
- 700g (1lb 8oz) lean pork
- 700g (1lb 8oz) fatty pork
- 170–230g (6–8oz) dried pinhead rusks or white breadcrumbs
- 110g (4oz) seasoning of your choice

Mince the meat, mix it all together then force it into the casing remembering to tie up the end before you start. You can have one long sausage, tie, or very carefully twist between each. Give your sausages 24 hours to rest and then cook them or freeze them – they have no preservative in them so will only keep for a limited time.

Glossary

AI – artificial insemination.

Colostrum – first milk the sow produces after birth. It is packed with antibodies.

Creep – an area in the sty that only small piglets can reach.

Creep pellets – high protein pellets fed to piglets from the age of three weeks.

Cross-breed – the offspring of two pure breeds mated together.

Crush – a type of metal cage used to confine an animal in order to administer drugs or artificial insemination.

Deadweight – the weight of a dressed carcass after slaughter.

Dished – an upturned snout.

Farrowing – giving birth.

Finishing – last stage of fattening before the pigs go to slaughter.

Gestation – period of pregnancy

Gilt – a female pig that has not produced a litter.

Hybrid – two pure breeds mated together to produce a cross.

In-pig – pregnant.

Lactation – the period during which a sow produces milk

Litter – a group of piglets from the same sow.

Liveweight – the total weight of pig when alive.

Lop-eared – a pig with ears that hang forward over its eyes.

Prick-eared – a pig with ears that stick straight up.

Runt – the smallest piglet in a litter

Scours/scouring – diarrhoea.

Slapboard – a board used for helping to move pigs.

Store pig – a pig being fattened between weaning and slaughter.

Terminal-sire – a breed of boar used to improve other breeds.

Weaner – an eight-week-old pig that has left its mother.

Useful websites

www.britishpigs.org.uk   **British Pig Association** – the home for information about all the UK's pedigree pig breeds including six at risk breeds. Contains independent industry information, details of shows and sales and club pages.

www.defra.gov.uk   **Department for the Environment, Food and Rural Affairs** – a comprehensive site covering all rules and regulations of every facet of farming.

www.rbst.org.uk   **The Rare Breeds Survival Trust** – a charity founded to protect all native breeds of farm animals, including poultry

www.soilassociation.org   **The Soil Association** – a membership charity that campaigns for planet-friendly organic food and farming.

Acknowledgements

I would particularly like to thank my sister Carol Keefer for her help and photographs of American breeds, Carl Blake for heaps of advice and help with American rare breeds, Roger Wakeling for butchery advice, my two sisters-in-law, Lucinda Roberts and Jo Seccombe for sharing their pig experiences (and the former for her poetic contribution), David Stocker, Helen Browning Organics, Steve Atkins of Pigs Paradise and the following people who either kindly sent me photos or let me use their pigs as models for the illustrations – and last but not least my long suffering husband for his poetic contribution and also allowing me to keep a few pigs in the interest of research.

Bayeux: Mike Lloyd.

Bentheim Black: Dr Jurgen Gunther Schulze, Guido Gerding, Beate Milerski.

Berkshire: Dawn Bull, Tracey and Graham Longhurst.

British Lop: Frank Miller, Steve Brooks.

British Saddleback: Sandy Helme and Sarah Harden, Tiggy and Angus Stovold.

Chester White: Cascade Meadows Farm, Adrian Brown, Carol Keefer.

Duroc: Jan Walton.

Gasçon: Roland Darre, Armin Grasse.

Gloucestershire Old Spots: Anne Nicholls.

Guinea Hog: Cascade Meadows Farm, Jessica Benson, Margaret Barkey.

Hampshire: Gary Cook, Sharon Barnfield.

Hereford: Lori Richardson.

Iron Age: Benjamin Weatherall, Cotswold Park Farm, Mark Boulton.

Kune Kune: Fran Waldron, Steve Atkins.

Lacombe: Canadian Swine Breeders Association.

Landrace: James Carne.

Large White: Mr Loveless, Helen Browning Organics.

Large Black: Gillian Dixon.

Limousin: Paul Morris.

Mangalitza: Siobhan Edwards.

Meishan: Carl Blake.

Middle White: Tracey and Graham Longhurst, Mandy Colbourne.

Mora Romagnola: Alessio Zanon.

Mulefoot: Carl Blake.

Ossabaw: Eliza Maclean, Carl Blake.

Oxford Sandy & Black: Tracey and Graham Longhurst.

Pietrain: Gillo Pedigree Pietrain Pigs, L. Makin.

Poland China: Carol Keefer.

Red Wattle: Dot Jordan.

Spotted: Carol Keefer.

Swabian Hall: Carl Blake, Beate Milerski.

Tamworth: Pauline Dixon, Dan Champion, Steve Atkins.

Welsh: Pedigree Welsh Pig Society.

Index